Everyone's Guide to Planet Mercury

Compiled by

Reatha Bain

Scribbles

Year of Publication 2018

ISBN : 9789352979493

Book Published by

Scribbles

(An Imprint of Alpha Editions)

email - alphaedis@gmail.com

Produced by: PediaPress GmbH
Limburg an der Lahn
Germany
http://pediapress.com/

Contents

Introduction

Mercury (planet)

\<indicator name="pp-default"\> 🔒 \</indicator\> \<indicator name="featured-star"\> ⭐ \</indicator\>

Mercury

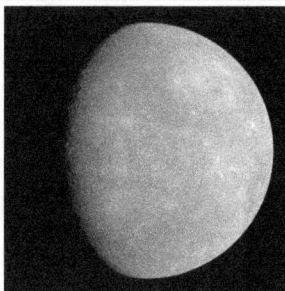

Mercury in enhanced color, imaged by *MESSENGER* (2008)

Designations	
Pronunciation	/ˈmɜːrkjʊri/ (🔊 listen)
Adjectives	Mercurian, mercurial
Orbital characteristics	
Epoch J2000	
Aphelion	• 0.466 697 AU • 69,816,900 km
Perihelion	• 0.307 499 AU • 46,001,200 km
Semi-major axis	• 0.387 098 AU • 57,909,050 km

Eccentricity	0.205 630
Orbital period	• 87.969 1 d • 0.240 846 yr • 0.5 Mercury solar day
Synodic period	115.88 d
Average orbital speed	47.362 km/s
Mean anomaly	174.796°
Inclination	• 7.005° to ecliptic • 3.38° to Sun's equator • 6.34° to invariable plane
Longitude of ascending node	48.331°
Argument of perihelion	29.124°
Satellites	None
Physical characteristics	
Mean radius	• 2,439.7±1.0 km • 0.3829 Earths
Flattening	0
Surface area	• 7.48×10^7 km^2 • 0.147 Earths
Volume	• 6.083×10^{10} km^3 • 0.056 Earths
Mass	• 3.3011×10^{23} kg • 0.055 Earths
Mean density	5.427 g/cm^3
Surface gravity	• 3.7 m/s^2 • 0.38 g
Moment of inertia factor	0.346±0.014
Escape velocity	4.25 km/s
Sidereal rotation period	• 58.646 d • 1407.5 h
Equatorial rotation velocity	10.892 km/h (3.026 m/s)
Axial tilt	2.04′ ± 0.08′ (to orbit) (0.034°)
North pole right ascension	• 18^h 44^m 2^s • 281.01°
North pole declination	61.45°

Albedo	• 0.068 (Bond) • 0.142 (geom.)				
		Surface temp.	**min**	**mean**	**max**
		0°N, 0°W	100 K	340 K	700 K
		85°N, 0°W	80 K	200 K	380 K
Apparent magnitude	−2.6 to 5.7				
Angular diameter	4.5–13"				
Atmosphere					
Surface pressure	trace (≤ 0.5 nPa)				
Composition by volume	• 42% molecular oxygen • 29.0% sodium • 22.0% hydrogen • 6.0% helium • 0.5% potassium • Trace amounts of argon, nitrogen, carbon dioxide, water vapor, xenon, krypton, and neon				

Mercury is the smallest and innermost planet in the Solar System. Its orbital period around the Sun of 87.97 days is the shortest of all the planets in the Solar System. It is named after the Roman deity Mercury, the messenger of the gods.

Like Venus, Mercury orbits the Sun within Earth's orbit as an *inferior planet*, and never exceeds 28° away from the Sun. When viewed from Earth, this proximity to the Sun means the planet can only be seen near the western or eastern horizon during the early evening or early morning. At this time it may appear as a bright star-like object, but is often far more difficult to observe than Venus. The planet telescopically displays the complete range of phases, similar to Venus and the Moon, as it moves in its inner orbit relative to Earth, which reoccurs over the so-called synodic period approximately every 116 days.

Mercury is tidally locked with the Sun in a 3:2 spin-orbit resonance,[1] and rotates in a way that is unique in the Solar System. As seen relative to the fixed stars, it rotates on its axis exactly three times for every two revolutions it makes around the Sun. As seen from the Sun, in a frame of reference that rotates with the orbital motion, it appears to rotate only once every two Mercurian years. An observer on Mercury would therefore see only one day every two years.

Mercury's axis has the smallest tilt of any of the Solar System's planets (about $1/30$ degree). Its orbital eccentricity is the largest of all known planets in the Solar System; at perihelion, Mercury's distance from the Sun is only about two-thirds (or 66%) of its distance at aphelion. Mercury's surface appears heavily

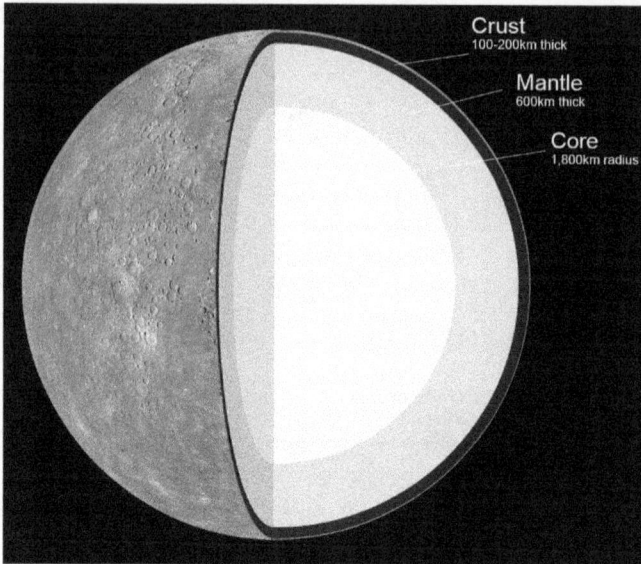

Figure 1:
Internal structure of Mercury:

1. *Crust: 100–300 km thick*
2. *Mantle: 600 km thick*
3. *Core: 1,800 km radius*

cratered and is similar in appearance to the Moon's, indicating that it has been geologically inactive for billions of years. Having almost no atmosphere to retain heat, it has surface temperatures that vary diurnally more than on any other planet in the Solar System, ranging from 100 K (–173 °C; –280 °F) at night to 700 K (427 °C; 800 °F) during the day across the equatorial regions. The polar regions are constantly below 180 K (–93 °C; –136 °F). The planet has no known natural satellites.

Two spacecraft have visited Mercury: *Mariner 10* flew by in 1974 and 1975; and *MESSENGER*, launched in 2004, orbited Mercury over 4,000 times in four years before exhausting its fuel and crashing into the planet's surface on April 30, 2015.

Physical characteristics

Internal structure

Mercury appears to have a solid silicate crust and mantle overlying a solid, iron sulfide outer core layer, a deeper liquid core layer, and possibly a solid inner

Figure 2: *Gravity anomalies on Mercury—mass concentrations (red) suggest subsurface structure and evolution*

core.

Mercury is one of four terrestrial planets in the Solar System, and is a rocky body like Earth. It is the smallest planet in the Solar System, with an equatorial radius of 2,439.7 kilometres (1,516.0 mi). Mercury is also smaller—albeit more massive—than the largest natural satellites in the Solar System, Ganymede and Titan. Mercury consists of approximately 70% metallic and 30% silicate material. Mercury's density is the second highest in the Solar System at 5.427 g/cm^3, only slightly less than Earth's density of 5.515 g/cm^3. If the effect of gravitational compression were to be factored out from both planets, the materials of which Mercury is made would be denser than those of Earth, with an uncompressed density of 5.3 g/cm^3 versus Earth's 4.4 g/cm^3.

Mercury's density can be used to infer details of its inner structure. Although Earth's high density results appreciably from gravitational compression, particularly at the core, Mercury is much smaller and its inner regions are not as compressed. Therefore, for it to have such a high density, its core must be large and rich in iron.

Geologists estimate that Mercury's core occupies about 55% of its volume; for Earth this proportion is 17%. Research published in 2007 suggests that Mercury has a molten core. Surrounding the core is a 500–700 km mantle consisting of silicates.[2] Based on data from the *Mariner 10* mission and Earth-based observation, Mercury's crust is estimated to be 35 km thick. One distinctive feature of Mercury's surface is the presence of numerous narrow ridges, extending up to several hundred kilometers in length. It is thought that these were formed as Mercury's core and mantle cooled and contracted at a time when the crust had already solidified.

Mercury's core has a higher iron content than that of any other major planet in the Solar System, and several theories have been proposed to explain this. The most widely accepted theory is that Mercury originally had a metal–silicate ratio similar to common chondrite meteorites, thought to be typical of the Solar System's rocky matter, and a mass approximately 2.25 times its current mass. Early in the Solar System's history, Mercury may have been struck by a planetesimal of approximately 1/6 that mass and several thousand kilometers across. The impact would have stripped away much of the original crust and mantle, leaving the core behind as a relatively major component. A similar process, known as the giant impact hypothesis, has been proposed to explain the formation of the Moon.

Alternatively, Mercury may have formed from the solar nebula before the Sun's energy output had stabilized. It would initially have had twice its present mass, but as the protosun contracted, temperatures near Mercury could have been between 2,500 and 3,500 K and possibly even as high as 10,000 K. Much of Mercury's surface rock could have been vaporized at such temperatures, forming an atmosphere of "rock vapor" that could have been carried away by the solar wind.

A third hypothesis proposes that the solar nebula caused drag on the particles from which Mercury was accreting, which meant that lighter particles were lost from the accreting material and not gathered by Mercury. Each hypothesis predicts a different surface composition, and there are two space missions set to make observations. *MESSENGER*, which ended in 2015, found higher-than-expected potassium and sulfur levels on the surface, suggesting that the giant impact hypothesis and vaporization of the crust and mantle did not occur because potassium and sulfur would have been driven off by the extreme heat of these events. *BepiColombo*, which will arrive at Mercury in 2025, will make observations to test these hypotheses. The findings so far would seem to favor the third hypothesis; however, further analysis of the data is needed.

Surface geology

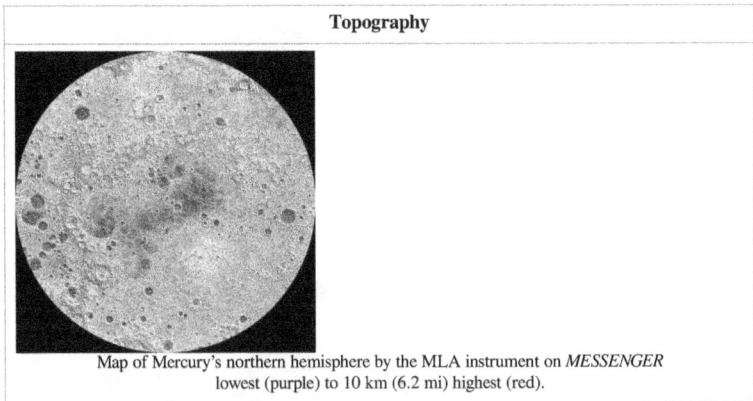

Topography

Map of Mercury's northern hemisphere by the MLA instrument on *MESSENGER* lowest (purple) to 10 km (6.2 mi) highest (red).

Mercury's surface is similar in appearance to that of the Moon, showing extensive mare-like plains and heavy cratering, indicating that it has been geologically inactive for billions of years. Because knowledge of Mercury's geology had been based only on the 1975 *Mariner 10* flyby and terrestrial observations, it is the least understood of the terrestrial planets. As data from *MESSENGER* orbiter are processed, this knowledge will increase. For example, an unusual crater with radiating troughs has been discovered that scientists called "the spider". It was later named Apollodorus.

Albedo features are areas of markedly different reflectivity, as seen by telescopic observation. Mercury has dorsa (also called "wrinkle-ridges"), Moon-like highlands, montes (mountains), planitiae (plains), rupes (escarpments), and valles (valleys).

Names for features on Mercury come from a variety of sources. Names coming from people are limited to the deceased. Craters are named for artists, musicians, painters, and authors who have made outstanding or fundamental contributions to their field. Ridges, or dorsa, are named for scientists who have contributed to the study of Mercury. Depressions or fossae are named for works of architecture. Montes are named for the word "hot" in a variety of languages. Plains or planitiae are named for Mercury in various languages. Escarpments or rupēs are named for ships of scientific expeditions. Valleys or valles are named for radio telescope facilities.

Mercury was heavily bombarded by comets and asteroids during and shortly following its formation 4.6 billion years ago, as well as during a possibly separate subsequent episode called the Late Heavy Bombardment that ended 3.8 billion years ago. During this period of intense crater formation, Mercury received impacts over its entire surface, facilitated by the lack of any atmosphere

Figure 3: *Mercury's surface*

Figure 4: *MASCS spectrum scan of Mercury's surface by MESSENGER*

Figure 5: *Perspective view of Caloris Basin – high (red); low (blue).*

to slow impactors down. During this time Mercury was volcanically active; basins such as the Caloris Basin were filled by magma, producing smooth plains similar to the maria found on the Moon.

Data from the October 2008 flyby of *MESSENGER* gave researchers a greater appreciation for the jumbled nature of Mercury's surface. Mercury's surface is more heterogeneous than either Mars's or the Moon's, both of which contain significant stretches of similar geology, such as maria and plateaus.

Impact basins and craters

Craters on Mercury range in diameter from small bowl-shaped cavities to multi-ringed impact basins hundreds of kilometers across. They appear in all states of degradation, from relatively fresh rayed craters to highly degraded crater remnants. Mercurian craters differ subtly from lunar craters in that the area blanketed by their ejecta is much smaller, a consequence of Mercury's stronger surface gravity. According to IAU rules, each new crater must be named after an artist that was famous for more than fifty years, and dead for more than three years, before the date the crater is named.

The largest known crater is Caloris Basin, with a diameter of 1,550 km. The impact that created the Caloris Basin was so powerful that it caused lava eruptions and left a concentric ring over 2 km tall surrounding the impact crater. At the antipode of the Caloris Basin is a large region of unusual, hilly terrain known as the "Weird Terrain". One hypothesis for its origin is that shock

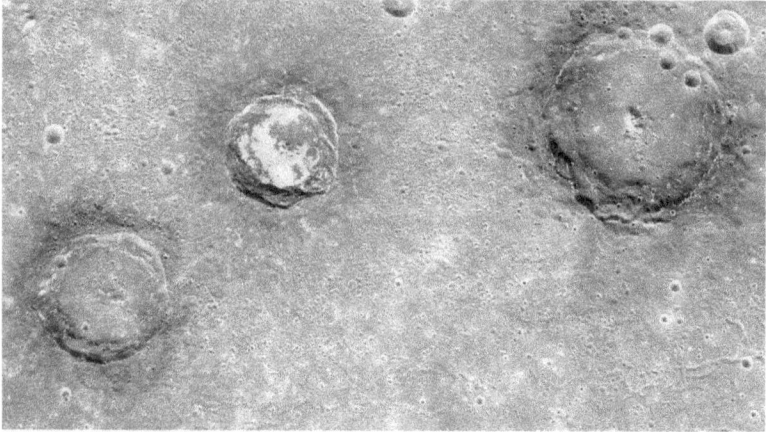

Figure 6: *Enhanced-color image of Munch, Sander and Poe craters amid volcanic plains (orange) near Caloris Basin*

waves generated during the Caloris impact traveled around Mercury, converging at the basin's antipode (180 degrees away). The resulting high stresses fractured the surface. Alternatively, it has been suggested that this terrain formed as a result of the convergence of ejecta at this basin's antipode.

Overall, about 15 impact basins have been identified on the imaged part of Mercury. A notable basin is the 400 km wide, multi-ring Tolstoj Basin that has an ejecta blanket extending up to 500 km from its rim and a floor that has been filled by smooth plains materials. Beethoven Basin has a similar-sized ejecta blanket and a 625 km diameter rim. Like the Moon, the surface of Mercury has likely incurred the effects of space weathering processes, including Solar wind and micrometeorite impacts.

Plains

There are two geologically distinct plains regions on Mercury. Gently rolling, hilly plains in the regions between craters are Mercury's oldest visible surfaces, predating the heavily cratered terrain. These inter-crater plains appear to have obliterated many earlier craters, and show a general paucity of smaller craters below about 30 km in diameter. <templatestyles src="Multiple_image/styles.css" />

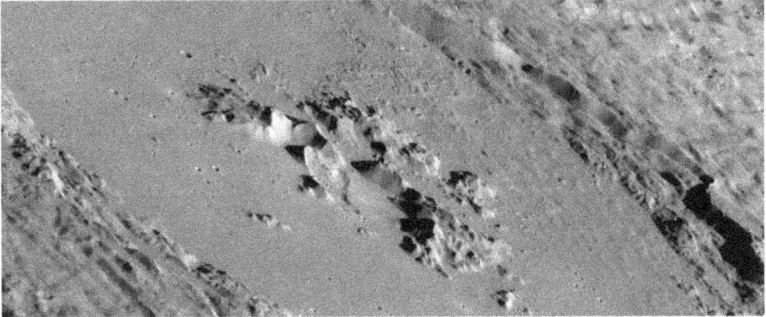

Figure 7:
Interior of Abedin crater

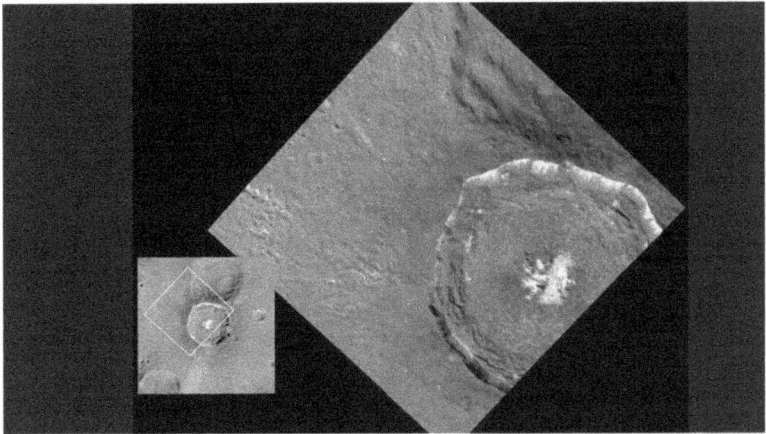

Figure 8: *Degas crater*

The so-called "Weird Terrain" formed at the point antipodal to the Caloris Basin impact

Smooth plains are widespread flat areas that fill depressions of various sizes and bear a strong resemblance to the lunar maria. Notably, they fill a wide ring surrounding the Caloris Basin. Unlike lunar maria, the smooth plains of Mercury have the same albedo as the older inter-crater plains. Despite a lack of unequivocally volcanic characteristics, the localisation and rounded, lobate shape of these plains strongly support volcanic origins. All the smooth plains of Mercury formed significantly later than the Caloris basin, as evidenced by appreciably smaller crater densities than on the Caloris ejecta blanket. The floor of the Caloris Basin is filled by a geologically distinct flat plain, broken up by ridges and fractures in a roughly polygonal pattern. It is not clear whether they are volcanic lavas induced by the impact, or a large sheet of impact melt.

Compressional features

One unusual feature of Mercury's surface is the numerous compression folds, or rupes, that crisscross the plains. As Mercury's interior cooled, it contracted and its surface began to deform, creating wrinkle ridges and lobate scarps associated with thrust faults. The scarps can reach lengths of 1000 km and heights of 3 km. These compressional features can be seen on top of other features, such as craters and smooth plains, indicating they are more recent. Mapping of the features has suggested a total shrinkage of Mercury's radius in the range of $\sim$1 to 7 km. Small-scale thrust fault scarps have been found, tens of meters in height and with lengths in the range of a few km, that appear to be less than 50 million years old, indicating that compression of the interior and consequent surface geological activity continue to the present.

The Lunar Reconnaissance Orbiter discovered that similar small thrust faults exist on the Moon.

Volcanology

Images obtained by *MESSENGER* have revealed evidence for pyroclastic flows on Mercury from low-profile shield volcanoes. *MESSENGER* data has helped identify 51 pyroclastic deposits on the surface, where 90% of them are found within impact craters. A study of the degradation state of the impact craters that host pyroclastic deposits suggests that pyroclastic activity occurred on Mercury over a prolonged interval.

A "rimless depression" inside the southwest rim of the Caloris Basin consists of at least nine overlapping volcanic vents, each individually up to 8 km in diameter. It is thus a "compound volcano". The vent floors are at a least 1 km below their brinks and they bear a closer resemblance to volcanic craters sculpted by explosive eruptions or modified by collapse into void spaces created by magma withdrawal back down into a conduit. The scientists could not quantify the age of the volcanic complex system, but reported that it could be of the order of a billion years.

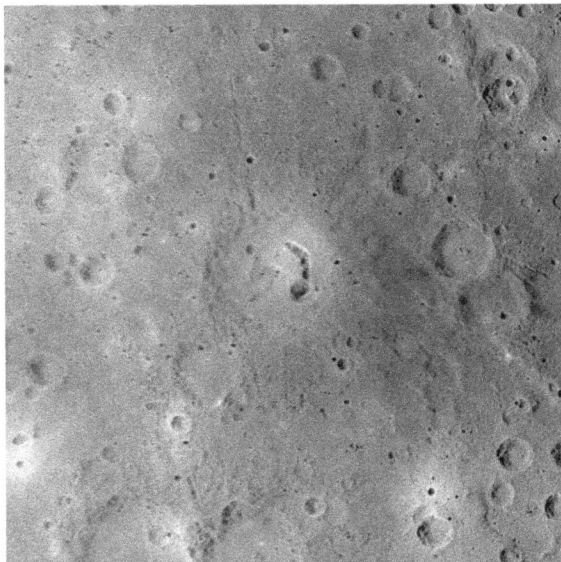

Figure 9: *Picasso crater — the large arc-shaped pit located on the eastern side of its floor are postulated to have formed when subsurface magma subsided or drained, causing the surface to collapse into the resulting void.*

Surface conditions and exosphere

The surface temperature of Mercury ranges from 100 to 700 K (–173 to 427 °C; –280 to 800 °F) at the most extreme places: 0°N, 0°W, or 180°W. It never rises above 180 K at the poles, due to the absence of an atmosphere and a steep temperature gradient between the equator and the poles. The subsolar point reaches about 700 K during perihelion (0°W or 180°W), but only 550 K at aphelion (90° or 270°W). On the dark side of the planet, temperatures average 110 K. The intensity of sunlight on Mercury's surface ranges between 4.59 and 10.61 times the solar constant (1,370 W·m^{-2}).

Although the daylight temperature at the surface of Mercury is generally extremely high, observations strongly suggest that ice (frozen water) exists on Mercury. The floors of deep craters at the poles are never exposed to direct sunlight, and temperatures there remain below 102 K; far lower than the global average. Water ice strongly reflects radar, and observations by the 70-meter Goldstone Solar System Radar and the VLA in the early 1990s revealed that there are patches of high radar reflection near the poles. Although ice was not the only possible cause of these reflective regions, astronomers think it was the most likely.

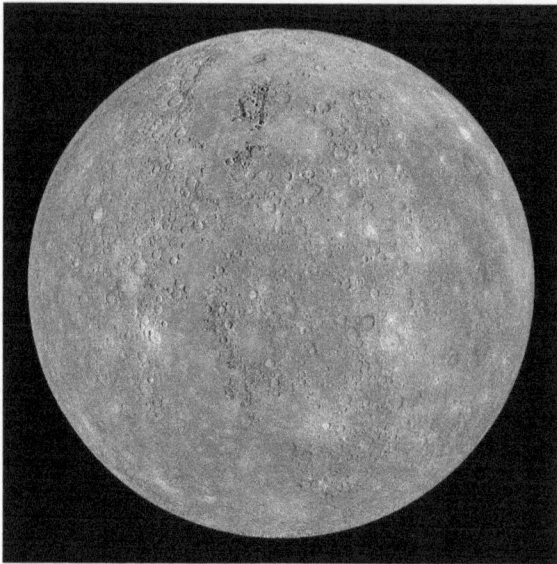

Figure 10: *Composite image of Mercury taken by MESSENGER*

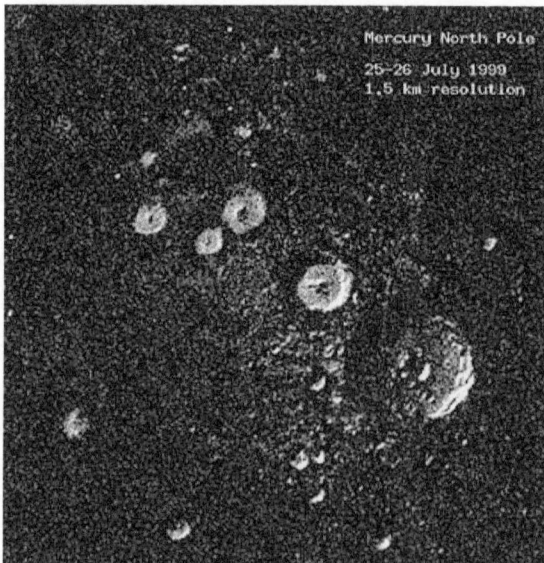

Figure 11: *Radar image of Mercury's north pole*

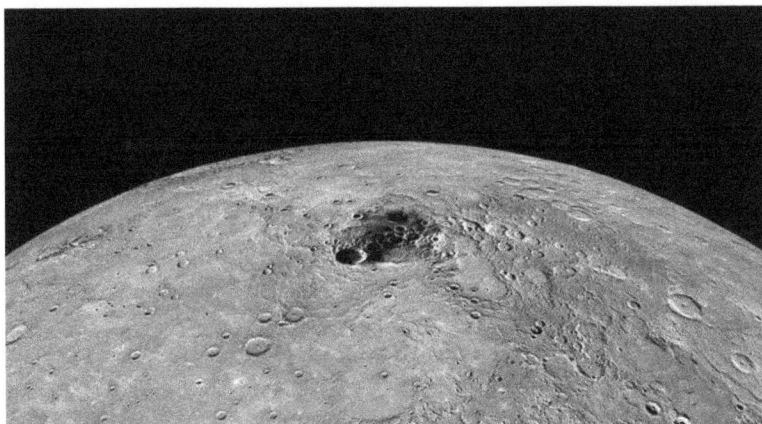

Figure 12: *Composite of the north pole of Mercury, where NASA confirmed the discovery of a large volume of water ice, in permanently dark craters that exist there.*

The icy regions are estimated to contain about 10^{14}–10^{15} kg of ice, and may be covered by a layer of regolith that inhibits sublimation. By comparison, the Antarctic ice sheet on Earth has a mass of about 4×10^{18} kg, and Mars's south polar cap contains about 10^{16} kg of water. The origin of the ice on Mercury is not yet known, but the two most likely sources are from outgassing of water from the planet's interior or deposition by impacts of comets.

Mercury is too small and hot for its gravity to retain any significant atmosphere over long periods of time; it does have a tenuous surface-bounded exosphere containing hydrogen, helium, oxygen, sodium, calcium, potassium and others at a surface pressure of less than approximately 0.5 nPa (0.005 picobars). This exosphere is not stable—atoms are continuously lost and replenished from a variety of sources. Hydrogen atoms and helium atoms probably come from the solar wind, diffusing into Mercury's magnetosphere before later escaping back into space. Radioactive decay of elements within Mercury's crust is another source of helium, as well as sodium and potassium. *MESSENGER* found high proportions of calcium, helium, hydroxide, magnesium, oxygen, potassium, silicon and sodium. Water vapor is present, released by a combination of processes such as: comets striking its surface, sputtering creating water out of hydrogen from the solar wind and oxygen from rock, and sublimation from reservoirs of water ice in the permanently shadowed polar craters. The detection of high amounts of water-related ions like O^+, OH^-, and H_2O^+ was a surprise. Because of the quantities of these ions that were detected in Mercury's space environment, scientists surmise that these molecules were blasted from the surface or exosphere by the solar wind.

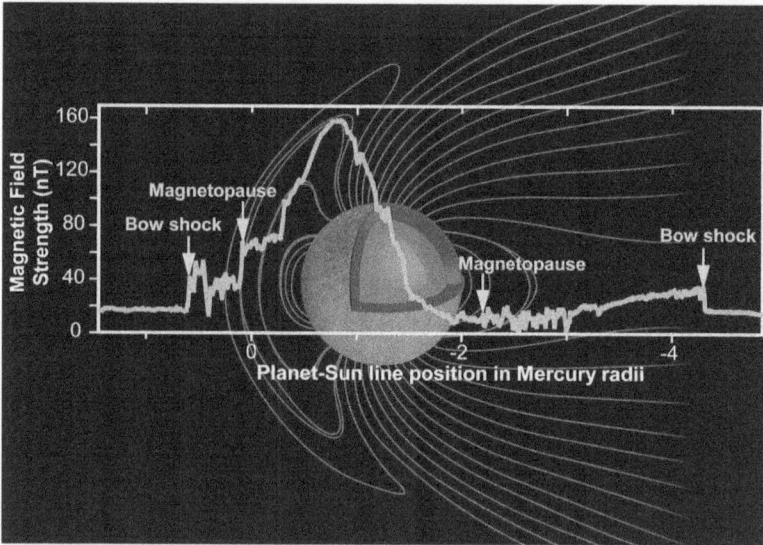

Figure 13: *Graph showing relative strength of Mercury's magnetic field*

Sodium, potassium and calcium were discovered in the atmosphere during the 1980–1990s, and are thought to result primarily from the vaporization of surface rock struck by micrometeorite impacts including presently from Comet Encke. In 2008, magnesium was discovered by *MESSENGER*. Studies indicate that, at times, sodium emissions are localized at points that correspond to the planet's magnetic poles. This would indicate an interaction between the magnetosphere and the planet's surface.

On November 29, 2012, NASA confirmed that images from *MESSENGER* had detected that craters at the north pole contained water ice. *MESSENGER*'s principal investigator Sean Solomon is quoted in *The New York Times* estimating the volume of the ice to be large enough to "encase Washington, D.C., in a frozen block two and a half miles deep".[3]

Magnetic field and magnetosphere

Despite its small size and slow 59-day-long rotation, Mercury has a significant, and apparently global, magnetic field. According to measurements taken by *Mariner 10*, it is about 1.1% the strength of Earth's. The magnetic-field strength at Mercury's equator is about 300 nT. Like that of Earth, Mercury's magnetic field is dipolar. Unlike Earth's, Mercury's poles are nearly aligned with the planet's spin axis. Measurements from both the *Mariner 10* and *MES-SENGER* space probes have indicated that the strength and shape of the magnetic field are stable.

It is likely that this magnetic field is generated by a dynamo effect, in a manner similar to the magnetic field of Earth. This dynamo effect would result from the circulation of the planet's iron-rich liquid core. Particularly strong tidal effects caused by the planet's high orbital eccentricity would serve to keep the core in the liquid state necessary for this dynamo effect.

Mercury's magnetic field is strong enough to deflect the solar wind around the planet, creating a magnetosphere. The planet's magnetosphere, though small enough to fit within Earth, is strong enough to trap solar wind plasma. This contributes to the space weathering of the planet's surface. Observations taken by the *Mariner 10* spacecraft detected this low energy plasma in the magnetosphere of the planet's nightside. Bursts of energetic particles in the planet's magnetotail indicate a dynamic quality to the planet's magnetosphere.

During its second flyby of the planet on October 6, 2008, *MESSENGER* discovered that Mercury's magnetic field can be extremely "leaky". The spacecraft encountered magnetic "tornadoes" – twisted bundles of magnetic fields connecting the planetary magnetic field to interplanetary space – that were up to 800 km wide or a third of the radius of the planet. These twisted magnetic flux tubes, technically known as flux transfer events, form open windows in the planet's magnetic shield through which the solar wind may enter and directly impact Mercury's surface via magnetic reconnection This also occurs in Earth's magnetic field. The *MESSENGER* observations showed the reconnection rate is ten times higher at Mercury, but its proximity to the Sun only accounts for about a third of the reconnection rate observed by *MESSENGER*.

Orbit, rotation, and longitude

<templatestyles src="Multiple_image/styles.css" />

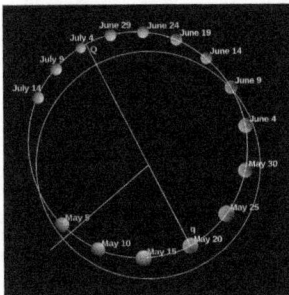

Orbit of Mercury (yellow). Dates refer to 2006.

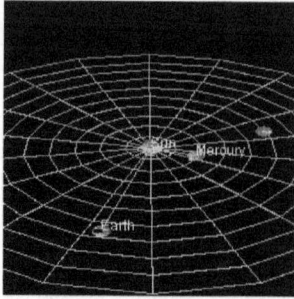

Animation of Mercury's and Earth's revolution around the Sun

Mercury has the most eccentric orbit of all the planets; its eccentricity is 0.21 with its distance from the Sun ranging from 46,000,000 to 70,000,000 km (29,000,000 to 43,000,000 mi). It takes 87.969 Earth days to complete an orbit. The diagram on the right illustrates the effects of the eccentricity, showing Mercury's orbit overlaid with a circular orbit having the same semi-major axis. Mercury's higher velocity when it is near perihelion is clear from the greater distance it covers in each 5-day interval. In the diagram the varying distance of Mercury to the Sun is represented by the size of the planet, which is inversely proportional to Mercury's distance from the Sun. This varying distance to the Sun leads to Mercury's surface being flexed by tidal bulges raised by the Sun that are about 17 times stronger than the Moon's on Earth. Combined with a 3:2 spin–orbit resonance of the planet's rotation around its axis, it also results in complex variations of the surface temperature. The resonance makes a single solar day on Mercury last exactly two Mercury years, or about 176 Earth days.

Mercury's orbit is inclined by 7 degrees to the plane of Earth's orbit (the ecliptic), as shown in the diagram on the right. As a result, transits of Mercury across the face of the Sun can only occur when the planet is crossing the plane of the ecliptic at the time it lies between Earth and the Sun. This occurs about every seven years on average.

Mercury's axial tilt is almost zero, with the best measured value as low as 0.027 degrees. This is significantly smaller than that of Jupiter, which has the second smallest axial tilt of all planets at 3.1 degrees. This means that to an observer at Mercury's poles, the center of the Sun never rises more than 2.1 arcminutes above the horizon.

At certain points on Mercury's surface, an observer would be able to see the Sun peek up about halfway over the horizon, then reverse and set before rising again, all within the same Mercurian day. This is because approximately four Earth days before perihelion, Mercury's angular orbital velocity equals its angular rotational velocity so that the Sun's apparent motion ceases; closer to perihelion, Mercury's angular orbital velocity then exceeds the angular rotational velocity. Thus, to a hypothetical observer on Mercury, the Sun appears to move in a retrograde direction. Four Earth days after perihelion, the Sun's normal apparent motion resumes. A similar effect would have occurred if Mercury had been in synchronous rotation: the alternating gain and loss of rotation over revolution would have caused a libration of 23.65° in longitude.

For the same reason, there are two points on Mercury's equator, 180 degrees apart in longitude, at either of which, around perihelion in alternate Mercurian years (once a Mercurian day), the Sun passes overhead, then reverses its apparent motion and passes overhead again, then reverses a second time and passes overhead a third time, taking a total of about 16 Earth-days for this entire process. In the other alternate Mercurian years, the same thing happens at the other of these two points. The amplitude of the retrograde motion is small, so the overall effect is that, for two or three weeks, the Sun is almost stationary overhead, and is at its most brilliant because Mercury is at perihelion, its closest to the Sun. This prolonged exposure to the Sun at its brightest makes these two points the hottest places on Mercury. Conversely, there are two other points on the equator, 90 degrees of longitude apart from the first ones, where the Sun passes overhead only when the planet is at aphelion in alternate years, when the apparent motion of the Sun in Mercury's sky is relatively rapid. These points, which are the ones on the equator where the apparent retrograde motion of the Sun happens when it is crossing the horizon as described in the preceding paragraph, receive much less solar heat than the first ones described above.

Mercury attains inferior conjunction (nearest approach to Earth) every 116 Earth days on average, but this interval can range from 105 days to 129 days due to the planet's eccentric orbit. Mercury can come as near as 82.2 gigametres (0.549 astronomical units; 51.1 million miles) to Earth, and that is slowly declining: The next approach to within 82.1 Gm (51.0 million miles) is in 2679, and to within 82.0 Gm (51.0 million miles) in 4487, but it will not be closer to Earth than 80 Gm (50 million miles) until 28,622.[4] Its period of retrograde motion as seen from Earth can vary from 8 to 15 days on either side of inferior conjunction. This large range arises from the planet's high orbital eccentricity.

Longitude convention

The longitude convention for Mercury puts the zero of longitude at one of the two hottest points on the surface, as described above. However, when this area was first visited, by *Mariner 10*, this zero meridian was in darkness, so it was impossible to select a feature on the surface to define the exact position of the meridian. Therefore, a small crater further west was chosen, called Hun Kal, which provides the exact reference point for measuring longitude.

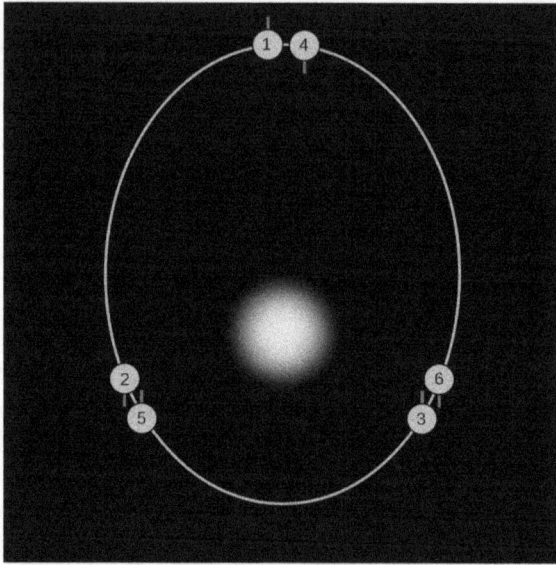

Figure 14: *After one orbit, Mercury has rotated 1.5 times, so after two complete orbits the same hemisphere is again illuminated.*

The center of Hun Kal defines the 20° West meridian. A 1970 International Astronomical Union resolution suggests that longitudes be measured positively in the westerly direction on Mercury. The two hottest places on the equator are therefore at longitudes 0°W and 180°W, and the coolest points on the equator are at longitudes 90°W and 270°W. However, the *MESSENGER* project uses an east-positive convention.

Spin–orbit resonance

For many years it was thought that Mercury was synchronously tidally locked with the Sun, rotating once for each orbit and always keeping the same face directed towards the Sun, in the same way that the same side of the Moon always faces Earth. Radar observations in 1965 proved that the planet has a 3:2 spin–orbit resonance, rotating three times for every two revolutions around the Sun. The eccentricity of Mercury's orbit makes this resonance stable—at perihelion, when the solar tide is strongest, the Sun is nearly still in Mercury's sky.

The rare 3:2 resonant tidal locking is stabilized by the variance of the tidal force along Mercury's eccentric orbit, acting on a permanent dipole component of Mercury's mass distribution. In a circular orbit there is no such variance, so the only resonance stabilized in such an orbit is at 1:1 (e.g., Earth–Moon), when the tidal force, stretching a body along the "center-body" line, exerts a torque that aligns the body's axis of least inertia (the "longest" axis, and the axis of the aforementioned dipole) to always point at the center. However, with noticeable eccentricity, like that of Mercury's orbit, the tidal force has a maximum at perihelion and thus stabilizes resonances, like 3:2, enforcing that the planet points its axis of least inertia roughly at the Sun when passing through perihelion.

The original reason astronomers thought it was synchronously locked was that, whenever Mercury was best placed for observation, it was always nearly at the same point in its 3:2 resonance, hence showing the same face. This is because, coincidentally, Mercury's rotation period is almost exactly half of its synodic period with respect to Earth. Due to Mercury's 3:2 spin–orbit resonance, a solar day (the length between two meridian transits of the Sun) lasts about 176 Earth days. A sidereal day (the period of rotation) lasts about 58.7 Earth days.

Simulations indicate that the orbital eccentricity of Mercury varies chaotically from nearly zero (circular) to more than 0.45 over millions of years due to perturbations from the other planets. This was thought to explain Mercury's 3:2 spin–orbit resonance (rather than the more usual 1:1), because this state is more likely to arise during a period of high eccentricity. However, accurate modeling based on a realistic model of tidal response has demonstrated that Mercury was captured into the 3:2 spin–orbit state at a very early stage of its history, within 20 (more likely, 10) million years after its formation.

Numerical simulations show that a future secular orbital resonant perihelion interaction with Jupiter may cause the eccentricity of Mercury's orbit to increase to the point where there is a 1% chance that the planet may collide with Venus within the next five billion years.

Advance of perihelion

In 1859, the French mathematician and astronomer Urbain Le Verrier reported that the slow precession of Mercury's orbit around the Sun could not be completely explained by Newtonian mechanics and perturbations by the known planets. He suggested, among possible explanations, that another planet (or perhaps instead a series of smaller 'corpuscles') might exist in an orbit even closer to the Sun than that of Mercury, to account for this perturbation.[5] (Other explanations considered included a slight oblateness of the Sun.) The success of the search for Neptune based on its perturbations of the orbit of Uranus led astronomers to place faith in this possible explanation, and the hypothetical planet was named Vulcan, but no such planet was ever found.

Figure 15: *Image mosaic by Mariner 10, 1974*

The perihelion precession of Mercury is 5,600 arcseconds (1.5556°) per century relative to Earth, or 574.10±0.65 arcseconds per century relative to the inertial ICRF. Newtonian mechanics, taking into account all the effects from the other planets, predicts a precession of 5,557 arcseconds (1.5436°) per century. In the early 20th century, Albert Einstein's general theory of relativity provided the explanation for the observed precession, by formalizing gravitation as being mediated by the curvature of spacetime. The effect is small: just 42.98 arcseconds per century for Mercury; it therefore requires a little over twelve million orbits for a full excess turn. Similar, but much smaller, effects exist for other Solar System bodies: 8.62 arcseconds per century for Venus, 3.84 for Earth, 1.35 for Mars, and 10.05 for 1566 Icarus.

Albert Einstein's formula for the perihelion shift per revolution is $\epsilon = 24\pi^3 \frac{a^2}{T^2 c^2 (1-e^2)}$, where e is the orbital eccentricity, a the semi-major axis, and T the orbital period. Filling in the values gives a result of 0.1035 arcseconds per revolution or 0.4297 arcseconds per Earth year, i.e., 42.97 arcseconds per century. This is in close agreement with the accepted value of Mercury's perihelion advance of 42.98 arcseconds per century.

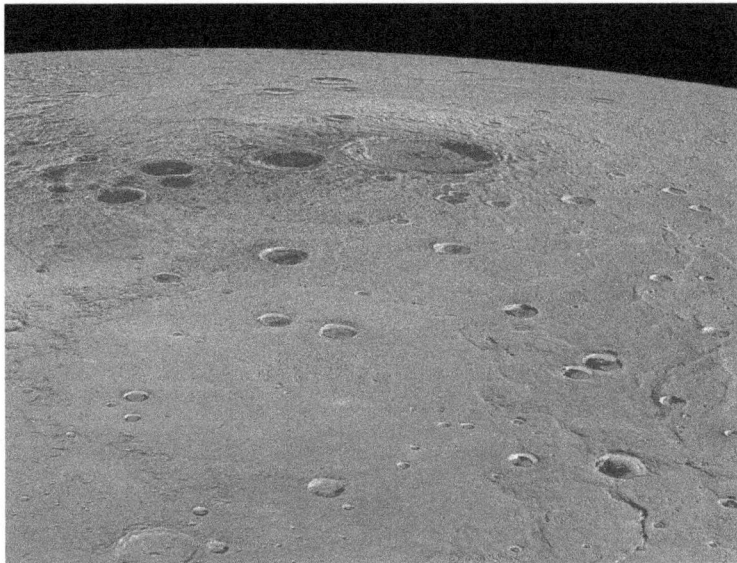

Figure 16: *False-color map showing the max-imum temperatures of the north polar region*

Observation

Mercury's apparent magnitude varies between –2.6 (brighter than the brightest star Sirius) and about +5.7 (approximating the theoretical limit of naked-eye visibility). The extremes occur when Mercury is close to the Sun in the sky. Observation of Mercury is complicated by its proximity to the Sun, as it is lost in the Sun's glare for much of the time. Mercury can be observed for only a brief period during either morning or evening twilight.

Mercury can, like several other planets and the brightest stars, be seen during a total solar eclipse.

Like the Moon and Venus, Mercury exhibits phases as seen from Earth. It is "new" at inferior conjunction and "full" at superior conjunction. The planet is rendered invisible from Earth on both of these occasions because of its being obscured by the Sun, except its new phase during a transit.

Mercury is technically brightest as seen from Earth when it is at a full phase. Although Mercury is farthest from Earth when it is full, the greater illuminated area that is visible and the opposition brightness surge more than compensates for the distance. The opposite is true for Venus, which appears brightest when it is a crescent, because it is much closer to Earth than when gibbous.

Figure 17: *False-color image of Carnegie Rupes, a tectonic landform—high terrain (red); low (blue).*

Nonetheless, the brightest (full phase) appearance of Mercury is an essentially impossible time for practical observation, because of the extreme proximity of the Sun. Mercury is best observed at the first and last quarter, although they are phases of lesser brightness. The first and last quarter phases occur at greatest elongation east and west of the Sun, respectively. At both of these times Mercury's separation from the Sun ranges anywhere from 17.9° at perihelion to 27.8° at aphelion.[6,7] At greatest *western* elongation, Mercury rises at its earliest before sunrise, and at greatest *eastern* elongation, it sets at its latest after sunset.

Mercury can be easily seen from the tropics and subtropics more than from higher latitudes. Viewed from low latitudes and at the right times of year, the ecliptic intersects the horizon at a steep angle. Mercury is 10° above the horizon when the planet appears directly above the Sun (i.e. its orbit appears vertical) and is at maximum elongation from the Sun (28°) and also when the Sun is 18° below the horizon, so the sky is just completely dark.[8] This angle is the maximum altitude at which Mercury is visible in a completely dark sky.

At middle latitudes, Mercury is more often and easily visible from the Southern Hemisphere than from the Northern. This is because Mercury's maximum western elongation occurs only during early autumn in the Southern Hemisphere, whereas its greatest eastern elongation happens only during late winter in the Southern Hemisphere. In both of these cases, the angle at which the planet's orbit intersects the horizon is maximized, allowing it to rise several hours before sunrise in the former instance and not set until several hours after sundown in the latter from southern mid-latitudes, such as Argentina and South Africa.

An alternate method for viewing Mercury involves observing the planet during daylight hours when conditions are clear, ideally when it is at its greatest elongation. This allows the planet to be found easily, even when using telescopes with 8 cm (3.1 in) apertures. Care must be taken to ensure the instrument isn't pointed directly towards the Sun because of the risk for eye damage. This method bypasses the limitation of twilight observing when the ecliptic is located at a low elevation (e.g. on autumn evenings).

Ground-based telescope observations of Mercury reveal only an illuminated partial disk with limited detail. The first of two spacecraft to visit the planet was *Mariner 10*, which mapped about 45% of its surface from 1974 to 1975. The second is the *MESSENGER* spacecraft, which after three Mercury flybys between 2008 and 2009, attained orbit around Mercury on March 17, 2011, to study and map the rest of the planet.

The Hubble Space Telescope cannot observe Mercury at all, due to safety procedures that prevent its pointing too close to the Sun.

Because the shift of 0.15 revolutions in a year makes up a seven-year cycle ($0.15 \times 7 \approx 1.0$), in the seventh year Mercury follows almost exactly (earlier by 7 days) the sequence of phenomena it showed seven years before.

Observation history

Ancient astronomers

The earliest known recorded observations of Mercury are from the Mul.Apin tablets. These observations were most likely made by an Assyrian astronomer around the 14th century BC. The cuneiform name used to designate Mercury on the Mul.Apin tablets is transcribed as Udu.Idim.Gu\u4.Ud ("the jumping planet"). Babylonian records of Mercury date back to the 1st millennium BC. The Babylonians called the planet Nabu after the messenger to the gods in their mythology.

The ancient Greeks knew the planet as Στίλβων (*Stilbon*), meaning "the gleaming", Ἑρμάων (*Hermaon*) and Ἑρμής (*Hermes*),[9] a planetary name that is retained in modern Greek (Ἑρμής: *Ermis*).[10] The Romans named the planet after the swift-footed Roman messenger god, Mercury (Latin *Mercurius*), which they equated with the Greek Hermes, because it moves across

Figure 18: *Mercury, from Liber astronomiae, 1550*

the sky faster than any other planet. The astronomical symbol for Mercury is a stylized version of Hermes' caduceus.

The Roman-Egyptian astronomer Ptolemy wrote about the possibility of planetary transits across the face of the Sun in his work *Planetary Hypotheses*. He suggested that no transits had been observed either because planets such as Mercury were too small to see, or because the transits were too infrequent.

In ancient China, Mercury was known as "the Hour Star" (*Chen-xing* 辰星). It was associated with the direction north and the phase of water in the Five Phases system of metaphysics. Modern Chinese, Korean, Japanese and Vietnamese cultures refer to the planet literally as the "water star" (水星), based on the Five elements. Hindu mythology used the name Budha for Mercury, and this god was thought to preside over Wednesday. The god Odin (or Woden) of Germanic paganism was associated with the planet Mercury and Wednesday. The Maya may have represented Mercury as an owl (or possibly four owls; two for the morning aspect and two for the evening) that served as a messenger to the underworld.

In medieval Islamic astronomy, the Andalusian astronomer Abū Ishāq Ibrāhīm al-Zarqālī in the 11th century described the deferent of Mercury's geocentric orbit as being oval, like an egg or a pignon, although this insight did not influence his astronomical theory or his astronomical calculations.[11] In the 12th century, Ibn Bajjah observed "two planets as black spots on the face of the Sun", which was later suggested as the transit of Mercury and/or Venus by the Maragha astronomer Qotb al-Din Shirazi in the 13th century. (Note that most such medieval reports of transits were later taken as observations of sunspots.)

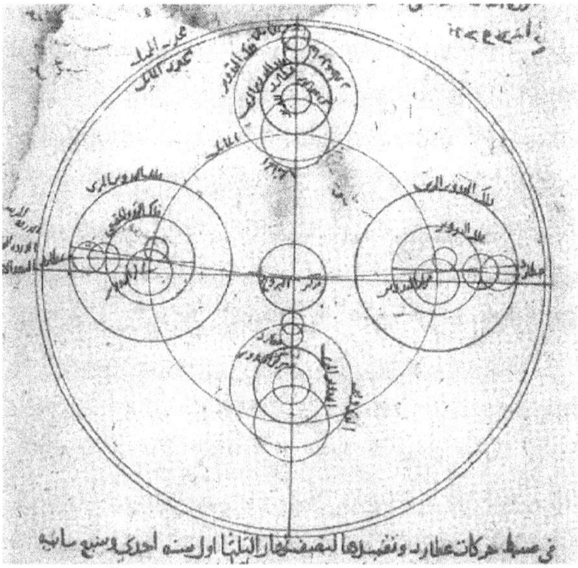

Figure 19: *Ibn al-Shatir's model for the appearances of Mercury, showing the multiplication of epicycles using the Tusi couple, thus eliminating the Ptolemaic eccentrics and equant.*

In India, the Kerala school astronomer Nilakantha Somayaji in the 15th century developed a partially heliocentric planetary model in which Mercury orbits the Sun, which in turn orbits Earth, similar to the Tychonic system later proposed by Tycho Brahe in the late 16th century.

Ground-based telescopic research

The first telescopic observations of Mercury were made by Galileo in the early 17th century. Although he observed phases when he looked at Venus, his telescope was not powerful enough to see the phases of Mercury. In 1631, Pierre Gassendi made the first telescopic observations of the transit of a planet across the Sun when he saw a transit of Mercury predicted by Johannes Kepler. In 1639, Giovanni Zupi used a telescope to discover that the planet had orbital phases similar to Venus and the Moon. The observation demonstrated conclusively that Mercury orbited around the Sun.

A rare event in astronomy is the passage of one planet in front of another (occultation), as seen from Earth. Mercury and Venus occult each other every few centuries, and the event of May 28, 1737 is the only one historically observed,

Figure 20: *Transit of Mercury. Mercury is visible as a black dot below and to the left of center. The dark area above the center of the solar disk is a sunspot.*

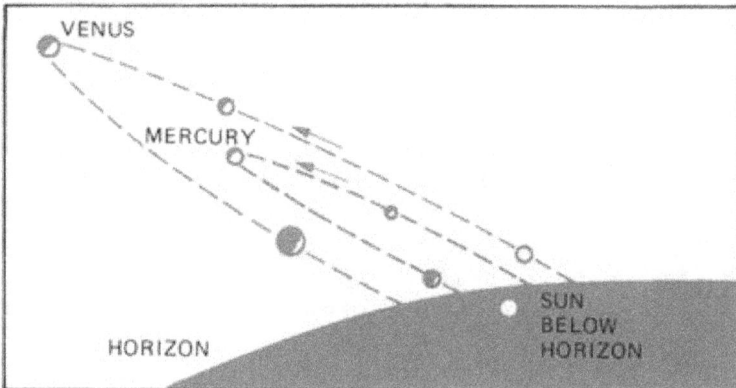

Figure 21: *Elongation is the angle between the Sun and the planet, with Earth as the reference point. Mercury appears close to the Sun.*

having been seen by John Bevis at the Royal Greenwich Observatory. The next occultation of Mercury by Venus will be on December 3, 2133.

The difficulties inherent in observing Mercury mean that it has been far less studied than the other planets. In 1800, Johann Schröter made observations of surface features, claiming to have observed 20-kilometre-high (12 mi) mountains. Friedrich Bessel used Schröter's drawings to erroneously estimate the rotation period as 24 hours and an axial tilt of 70°. In the 1880s, Giovanni Schiaparelli mapped the planet more accurately, and suggested that Mercury's rotational period was 88 days, the same as its orbital period due to tidal locking. This phenomenon is known as synchronous rotation. The effort to map the surface of Mercury was continued by Eugenios Antoniadi, who published a book in 1934 that included both maps and his own observations. Many of the planet's surface features, particularly the albedo features, take their names from Antoniadi's map.

In June 1962, Soviet scientists at the Institute of Radio-engineering and Electronics of the USSR Academy of Sciences, led by Vladimir Kotelnikov, became the first to bounce a radar signal off Mercury and receive it, starting radar observations of the planet. Three years later, radar observations by Americans Gordon Pettengill and R. Dyce, using the 300-meter Arecibo Observatory radio telescope in Puerto Rico, showed conclusively that the planet's rotational period was about 59 days.[12] The theory that Mercury's rotation was synchronous had become widely held, and it was a surprise to astronomers when these radio observations were announced. If Mercury were tidally locked, its dark face would be extremely cold, but measurements of radio emission revealed that it was much hotter than expected. Astronomers were reluctant to drop the synchronous rotation theory and proposed alternative mechanisms such as powerful heat-distributing winds to explain the observations.

Italian astronomer Giuseppe Colombo noted that the rotation value was about two-thirds of Mercury's orbital period, and proposed that the planet's orbital and rotational periods were locked into a 3:2 rather than a 1:1 resonance. Data from *Mariner 10* subsequently confirmed this view. This means that Schiaparelli's and Antoniadi's maps were not "wrong". Instead, the astronomers saw the same features during every *second* orbit and recorded them, but disregarded those seen in the meantime, when Mercury's other face was toward the Sun, because the orbital geometry meant that these observations were made under poor viewing conditions.

Ground-based optical observations did not shed much further light on Mercury, but radio astronomers using interferometry at microwave wavelengths, a technique that enables removal of the solar radiation, were able to discern physical and chemical characteristics of the subsurface layers to a depth of several meters.[13] Not until the first space probe flew past Mercury did many

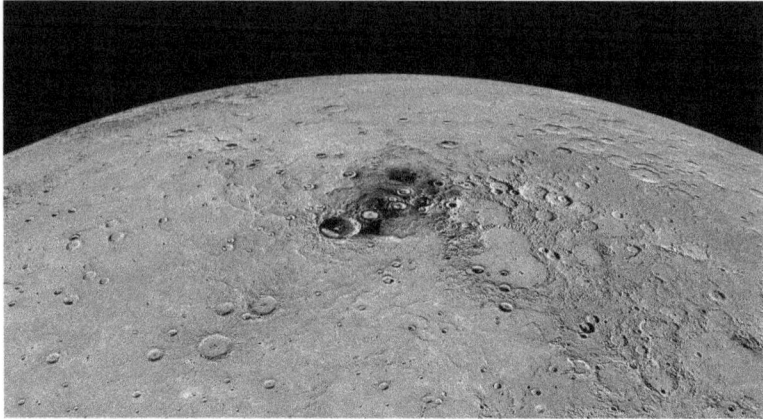

Figure 22: *Water ice (yellow) at Mercury's north polar region*

of its most fundamental morphological properties become known. Moreover, recent technological advances have led to improved ground-based observations. In 2000, high-resolution lucky imaging observations were conducted by the Mount Wilson Observatory 1.5 meter Hale telescope. They provided the first views that resolved surface features on the parts of Mercury that were not imaged in the *Mariner 10* mission. Most of the planet has been mapped by the Arecibo radar telescope, with 5 km (3.1 mi) resolution, including polar deposits in shadowed craters of what may be water ice.

Research with space probes

Reaching Mercury from Earth poses significant technical challenges, because it orbits so much closer to the Sun than Earth. A Mercury-bound spacecraft launched from Earth must travel over 91 million kilometres (57 million miles) into the Sun's gravitational potential well. Mercury has an orbital speed of 48 km/s (30 mi/s), whereas Earth's orbital speed is 30 km/s (19 mi/s). Therefore, the spacecraft must make a large change in velocity (delta-v) to enter a Hohmann transfer orbit that passes near Mercury, as compared to the delta-v required for other planetary missions.

The potential energy liberated by moving down the Sun's potential well becomes kinetic energy; requiring another large delta-v change to do anything other than rapidly pass by Mercury. To land safely or enter a stable orbit the spacecraft would rely entirely on rocket motors. Aerobraking is ruled out because Mercury has a negligible atmosphere. A trip to Mercury requires more rocket fuel than that required to escape the Solar System completely. As a result, only two space probes have visited it so far. A proposed alternative

Figure 23: *MESSENGER being prepared for launch*

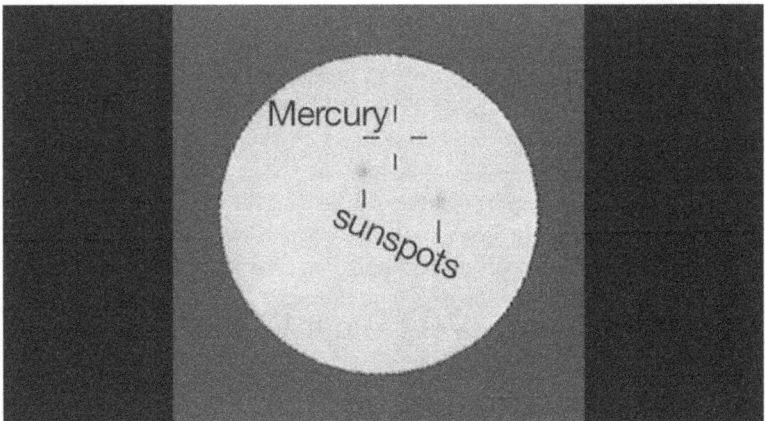

Figure 24: *Mercury transiting the Sun as viewed by the Mars rover Curiosity (June 3, 2014).*

Figure 25: *Mariner 10, the first probe to visit Mercury*

approach would use a solar sail to attain a Mercury-synchronous orbit around the Sun.

Mariner 10

The first spacecraft to visit Mercury was NASA's *Mariner 10* (1974–1975). The spacecraft used the gravity of Venus to adjust its orbital velocity so that it could approach Mercury, making it both the first spacecraft to use this gravitational "slingshot" effect and the first NASA mission to visit multiple planets. *Mariner 10* provided the first close-up images of Mercury's surface, which immediately showed its heavily cratered nature, and revealed many other types of geological features, such as the giant scarps that were later ascribed to the effect of the planet shrinking slightly as its iron core cools. Unfortunately, the same face of the planet was lit at each of *Mariner 10*'s close approaches. This made close observation of both sides of the planet impossible, and resulted in the mapping of less than 45% of the planet's surface.

The spacecraft made three close approaches to Mercury, the closest of which took it to within 327 km (203 mi) of the surface. At the first close approach, instruments detected a magnetic field, to the great surprise of planetary geologists—Mercury's rotation was expected to be much too slow to generate a significant dynamo effect. The second close approach was primarily used for imaging, but at the third approach, extensive magnetic data were obtained.

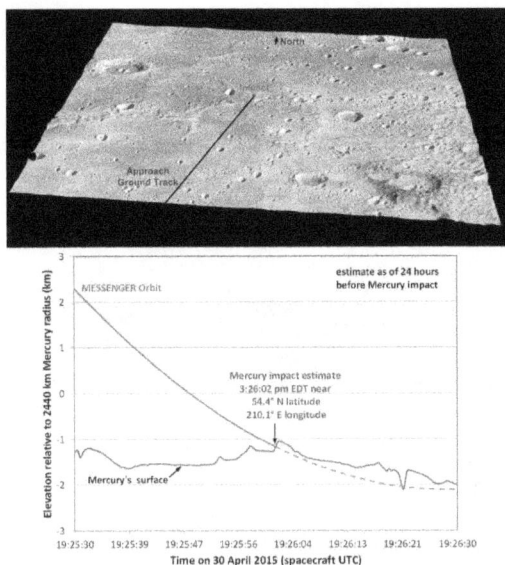

Figure 26: *Estimated details of the impact of MESSENGER on April 30, 2015*

The data revealed that the planet's magnetic field is much like Earth's, which deflects the solar wind around the planet. For many years after the *Mariner 10* encounters, the origin of Mercury's magnetic field remained the subject of several competing theories.

On March 24, 1975, just eight days after its final close approach, *Mariner 10* ran out of fuel. Because its orbit could no longer be accurately controlled, mission controllers instructed the probe to shut down. *Mariner 10* is thought to be still orbiting the Sun, passing close to Mercury every few months.

MESSENGER

A second NASA mission to Mercury, named *MESSENGER* (MErcury Surface, Space ENvironment, GEochemistry, and Ranging), was launched on August 3, 2004. It made a fly-by of Earth in August 2005, and of Venus in October 2006 and June 2007 to place it onto the correct trajectory to reach an orbit around Mercury. A first fly-by of Mercury occurred on January 14, 2008, a second on October 6, 2008, and a third on September 29, 2009. Most of the hemisphere not imaged by *Mariner 10* was mapped during these fly-bys. The probe successfully entered an elliptical orbit around the planet on March 18, 2011. The first orbital image of Mercury was obtained on March 29, 2011. The probe finished a one-year mapping mission, and then entered

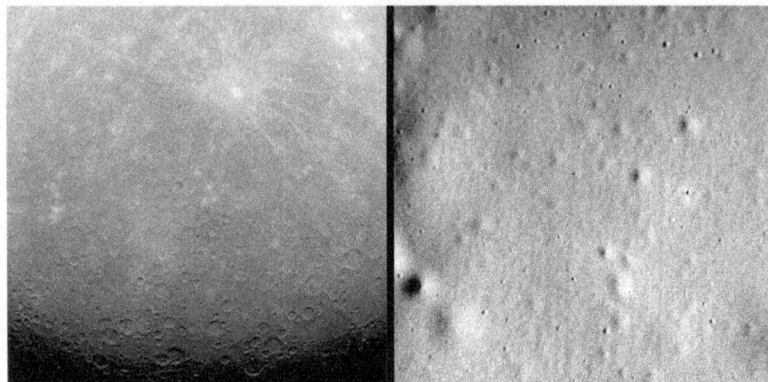

Figure 27: *First (March 29, 2011) and last (April 30, 2015) images of Mercury by MESSENGER*

a one-year extended mission into 2013. In addition to continued observations and mapping of Mercury, *MESSENGER* observed the 2012 solar maximum.[14]

The mission was designed to clear up six key issues: Mercury's high density, its geological history, the nature of its magnetic field, the structure of its core, whether it has ice at its poles, and where its tenuous atmosphere comes from. To this end, the probe carried imaging devices that gathered much-higher-resolution images of much more of Mercury than *Mariner 10*, assorted spectrometers to determine abundances of elements in the crust, and magnetometers and devices to measure velocities of charged particles. Measurements of changes in the probe's orbital velocity were expected to be used to infer details of the planet's interior structure. *MESSENGER*'s final maneuver was on April 24, 2015, and it crashed into Mercury's surface on April 30, 2015. The spacecraft's impact with Mercury occurred near 3:26 PM EDT on April 30, 2015, leaving a crater estimated to be 16 m (52 ft) in diameter.

BepiColombo

The European Space Agency and the Japanese Space Agency are planning a joint mission called *BepiColombo*, which will orbit Mercury with two probes: one to map the planet and the other to study its magnetosphere. Once launched in 2018, *BepiColombo* is expected to reach Mercury in 2025. It will release a magnetometer probe into an elliptical orbit, then chemical rockets will fire to deposit the mapper probe into a circular orbit. Both probes will operate for one terrestrial year. The mapper probe will carry an array of spectrometers similar to those on *MESSENGER*, and will study the planet at many different wavelengths including infrared, ultraviolet, X-ray and gamma ray.

Comparison

<templatestyles src="Multiple_image/styles.css" />

Size comparison with other Solar System objects

Mercury, Earth

Mercury, Venus, Earth, Mars

Back row: Mars, Mercury
Front: Moon, Pluto, Haumea

External links

<indicator name="spoken-icon"> ◉)) </indicator>

- *Atlas of Mercury*[15]. NASA. 1978. SP-423.
- Mercury nomenclature[16] and map with feature names[17] from the USGS/ IAU *Gazetteer of Planetary Nomenclature*
- Equirectangular map of Mercury[18] by Applied Coherent Technology Corp
- 3D globe of Mercury[19] by Google
- Mercury[20] at Solarviews.com
- Mercury[21] by Astronomy Cast
- *MESSENGER* mission web site[22]
- *BepiColombo* mission web site[23]

Geology

Geology of Mercury

The **geology of Mercury** is the least understood of all the terrestrial planets in the Solar System. This stems largely from Mercury's proximity to the Sun which makes reaching it with spacecraft technically challenging and Earth-based observations difficult.

Mercury's surface is dominated by impact craters, basaltic rock and smooth plains, many of them a result of flood volcanism, similar in some respects to the lunar maria, and locally by pyroclastic deposits. Other notable features include vents which appear to be the source of magma-carved valleys, often-grouped irregular-shaped depressions termed "hollows" that are believed to be the result of collapsed magma chambers, scarps indicative of thrust faulting and mineral deposits (possibly ice) inside craters at the poles. Long thought to be geologically inactive, new evidence suggests there may still be some level of activity.

Mercury's density implies a solid iron-rich core that accounts for about 60% of its volume (75% of its radius). Mercury's magnetic equator is shifted nearly 20% of the planet's radius towards the north, the largest ratio of all planets. This shift lends to there being one or more iron-rich molten layers surrounding the core producing a dynamo effect similar to that of Earth. Additionally, the offset magnetic dipole may result in uneven surface weathering by the solar wind, knocking more surface particles up into the southern exosphere and transporting them for deposit in the north. Scientists are gathering telemetry to determine if such is the case.

After having completed the first solar day of its mission in September 2011, more than 99% of Mercury's surface had been mapped by NASA's *MESSENGER* probe in both color and monochrome with such detail that scientists' understanding of Mercury's geology has eclipsed the level achieved following the *Mariner 10* flybys of the 1970s.

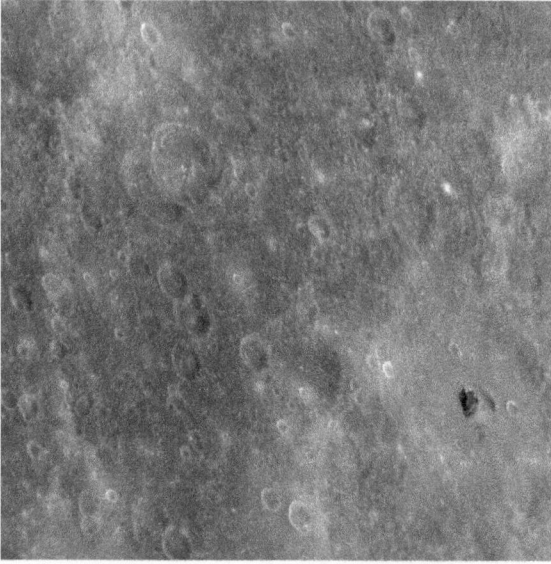

Figure 28: *An unexplained patch of black on Mercury.*

Figure 29: *A double-ring impact basin on Mercury.*

Figure 30: *Mariner 10 probe*

Difficulties in exploration

Reaching Mercury from Earth poses significant technical challenges, because the planet orbits so much closer to the Sun than does the Earth. A Mercury-bound spacecraft launched from Earth must travel 91 million kilometers into the Sun's gravitational potential well. Starting from the Earth's orbital speed of 30 km/s, the change in velocity (delta-v) the spacecraft must make to enter into a Hohmann transfer orbit that passes near Mercury is large compared to other planetary missions. The potential energy liberated by moving down the Sun's potential well becomes kinetic energy; requiring another large delta-v to do anything other than rapidly pass by Mercury. In order to land safely or enter a stable orbit the spacecraft must rely entirely on rocket motors because Mercury has negligible atmosphere. A direct trip to Mercury actually requires more rocket fuel than that required to escape the Solar System completely. As a result, only two space probes, *Mariner 10* and *MESSENGER*, both by NASA, have visited Mercury so far.

Furthermore, the space environment near Mercury is demanding, posing the double dangers to spacecraft of intense solar radiation and high temperatures.

Historically, a second obstacle has been that Mercury's period of rotation is a slow 58 Earth days, so that spacecraft flybys are restricted to viewing only a single illuminated hemisphere. In fact, unfortunately, even though Mariner 10

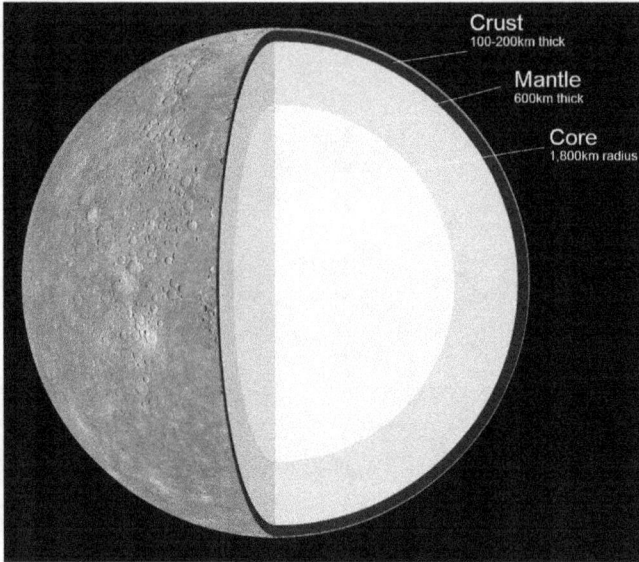

Figure 31: *1. Crust – 100–200 km thick*
2. Mantle – 600 km thick
3. Nucleus – 1,800 km radius

space probe flew past Mercury three times during 1974 and 1975, it observed the same area during each pass. This was because Mariner 10's orbital period was almost exactly 3 sidereal Mercury days, and the same face of the planet was lit at each of the close approaches. As a result, less than 45% of the planet's surface was mapped.

Earth-based observations are made difficult by Mercury's constant proximity to the Sun. This has several consequences:

1. Whenever the sky is dark enough for viewing through telescopes, Mercury is always already near the horizon, where viewing conditions are poor anyway due to atmospheric factors.
2. The Hubble Space Telescope and other space observatories are usually prevented from pointing close to the Sun for safety reasons (Erroneously pointing such sensitive instruments at the Sun is likely to cause permanent damage).

Figure 32: *Mercury – Gravity Anomalies – mass concentrations (red) suggest subsurface structure and evolution.*

Mercury's geological history

Like the Earth, Moon and Mars, Mercury's geologic history is divided up into eras. From oldest to youngest, these are: the pre-Tolstojan, Tolstojan, Calorian, Mansurian, and Kuiperian. These ages are based on relative dating only.

After the formation of Mercury along with the rest of the Solar System 4.6 billion years ago, heavy bombardment by asteroids and comets ensued. The last intense bombardment phase, the Late Heavy Bombardment came to an end about 3.8 billion years ago. Some regions or massifs, a prominent one being the one that formed the Caloris Basin, were filled by magma eruptions from within the planet. These created smooth intercrater plains similar to the maria found on the Moon. Later, as the planet cooled and contracted, its surface began to crack and form ridges; these surface cracks and ridges can be seen on top of other features, such as the craters and smoother plains—a clear indication that they are more recent. Mercury's period of volcanism ended when the planet's mantle had contracted enough to prevent further lava from breaking through to the surface. This probably occurred at some point during its first 700 or 800 million years of history.

Since then, the main surface processes have been intermittent impacts.

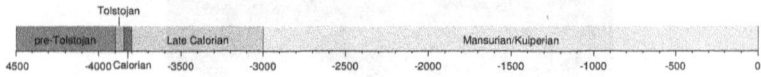

Tolstojan								
pre-Tolstojan	Late Calorian				Mansurian/Kuiperian			
4500	-4000 Calorian -3500	-3000	-2500	-2000	-1500	-1000	-500	0

Timeline

Time unit: millions of years

Surface features

> Wikinews has related news: *NASA releases first topographical map of Mercury*

Mercury's surface is overall similar in appearance to that of the Moon, with extensive mare-like plains and heavily cratered terrains similar to the lunar highlands and made locally by accumulations of pyroclastic deposits.

Topography
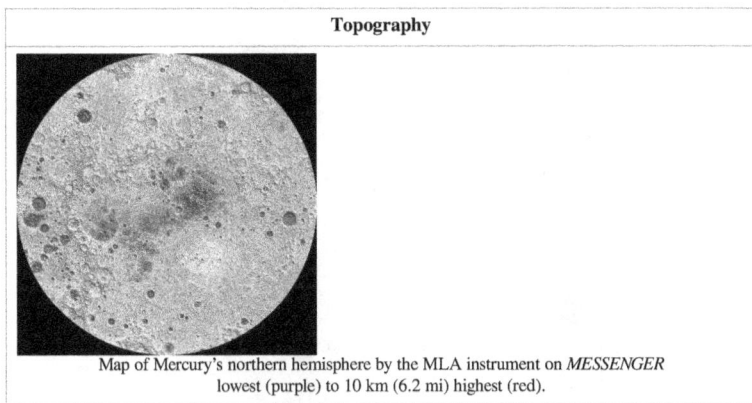
Map of Mercury's northern hemisphere by the MLA instrument on *MESSENGER* lowest (purple) to 10 km (6.2 mi) highest (red).

Impact basins and craters

Craters on Mercury range in diameter from small bowl-shaped craters to multi-ringed impact basins hundreds of kilometers across. They appear in all states of degradation, from relatively fresh rayed-craters, to highly degraded crater remnants. Mercurian craters differ subtly from Lunar craters — the extent of their ejecta blankets is much smaller, which is a consequence of the 2.5 times stronger surface gravity on Mercury.

The largest known crater is the enormous Caloris Basin, with a diameter of 1550 km, A basin of comparable size, tentatively named Skinakas Basin had been postulated from low resolution Earth-based observations of the non-Mariner-imaged hemisphere, but has not been observed in *MESSENGER* imagery of the corresponding terrain. The impact which created the Caloris Basin was so powerful that its effects are seen on a global scale. It caused lava eruptions and left a concentric ring over 2 km tall surrounding the impact crater. At the antipode of the Caloris Basin lies a large region of unusual, hilly and furrowed terrain, sometimes called "Weird Terrain". The favoured hypothesis for the origin of this geomorphologic unit is that shock waves generated during the impact traveled around the planet, and when they converged at the basin's antipode (180 degrees away) the high stresses were capable of fracturing the surface. A much less favoured idea was that this terrain formed as a result of the convergence of ejecta at this basin's antipode. Furthermore, the formation of the Caloris Basin appears to have produced a shallow depression concentric around the basin, which was later filled by the smooth plains (see below).

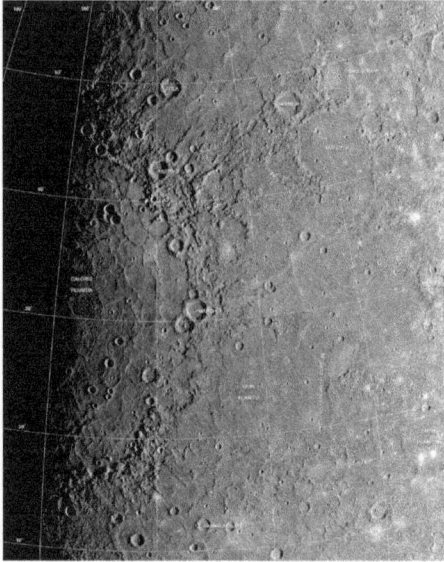

Figure 33: *Mercury's Caloris Basin is one of the largest impact features in the Solar System*

Figure 34: *MASCS spectrum scan of Mercury's surface by MESSENGER*

Figure 35: *The so-called "Weird Terrain" formed
by the Caloris Basin impact at its antipodal point*

Overall about 15 impact basins have been identified on the imaged part of Mercury. Other notable basins include the 400 km wide, multi-ring, Tolstoj Basin which has an ejecta blanket extending up to 500 km from its rim, and its floor has been filled by smooth plains materials. Beethoven Basin also has a similar-sized ejecta blanket and a 625 km diameter rim.

As on the Moon, fresh craters on Mercury show prominent bright ray systems. These are made by ejected debris, which tend to be brighter while they remain relatively fresh because of a lesser amount of space weathering than the surrounding older terrain.

Pit-floor craters

Some impact craters on Mercury have non-circular, irregularly shaped depressions or pits on their floors. Such craters have been termed pit-floor craters, and *MESSENGER* team members have suggested that such pits formed by the collapse of subsurface magma chambers. If this suggestion is correct, the pits are evidence of volcanic processes at work on Mercury. Pit craters are rimless, often irregularly shaped, and steep-sided, and they display no associated ejecta or lava flows but are typically distinctive in color. For example, the pits of Praxiteles have an orange hue. Thought to be evidence of shallow magmatic activity, pit craters may have formed when subsurface magma drained

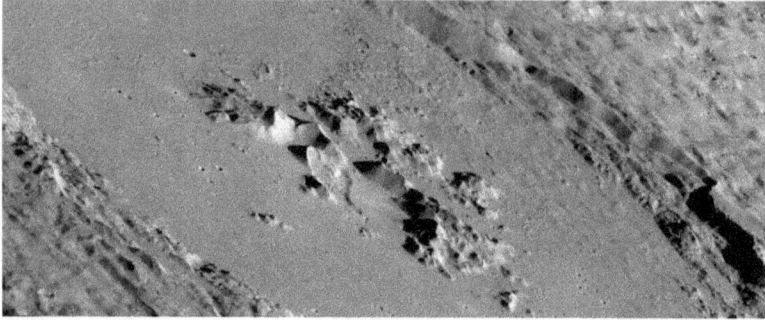

Figure 36:
Interior of Abedin crater

elsewhere and left a roof area unsupported, leading to collapse and the formation of the pit. Major craters exhibiting these features include Beckett, Gibran
and Lermontov, among others. It was suggested that these pits with associated
brighter and redder deposits may be pyroclastic deposits caused by explosive
volcanism.

Plains

There are two geologically distinct plains units on Mercury:

* *Inter-crater plains* are the oldest visible surface, predating the heavily
  cratered terrain. They are gently rolling or hilly and occur in the regions
  between larger craters. The inter-crater plains appear to have obliterated
  many earlier craters, and show a general paucity of smaller craters below
  about 30 km in diameter. It is not clear whether they are of volcanic or
  impact origin. The inter-crater plains are distributed roughly uniformly
  over the entire surface of the planet.
* *Smooth plains* are widespread flat areas resembling the lunar maria, which
  fill depressions of various sizes. Notably, they fill a wide ring surrounding
  the Caloris Basin. An appreciable difference to the lunar maria is that the
  smooth plains of Mercury have the same albedo as the older intercrater
  plains. Despite a lack of unequivocally volcanic features, their localisation
  and lobate-shaped colour units strongly support a volcanic origin. All the
  Mercurian smooth plains formed significantly later than the Caloris basin,
  as evidenced by appreciably smaller crater densities than on the Caloris
  ejecta blanket.

The floor of the Caloris Basin is also filled by a geologically distinct flat plain,
broken up by ridges and fractures in a roughly polygonal pattern. It is not

Figure 37: *Discovery Rupes.*

clear whether they are volcanic lavas induced by the impact, or a large sheet
of impact melt.

Tectonic features

One unusual feature of the planet's surface is the numerous compression folds
which crisscross the plains. It is thought that as the planet's interior cooled, it
contracted and its surface began to deform. The folds can be seen on top of
other features, such as craters and smoother plains, indicating that they are
more recent.[24] Mercury's surface is also flexed by significant tidal bulges
raised by the Sun—the Sun's tides on Mercury are about 17% stronger than
the Moon's on Earth.[25]

Terminology

Non-crater surface features are given the following names:

- Albedo features — areas of markedly different reflectivity
- Dorsa — ridges (*see List of ridges on Mercury*)
- Montes — mountains (*see List of mountains on Mercury*)
- Planitiae — plains (*see List of plains on Mercury*)
- Rupes — scarps (*see List of scarps on Mercury*)
- Valles — valleys (*see List of valleys on Mercury*)

Figure 38: *Radar image of Mercury's north pole.*

High-albedo polar patches and possible presence of ice

The first radar observations of Mercury were carried out by the radiotelescopes at Arecibo (Puerto Rico) and Goldstone (California, United States), with assistance from the U.S. National Radio Astronomy Observatory Very Large Array (VLA) facility in New Mexico. The transmissions sent from the NASA Deep Space Network site at Goldstone were at a power level of 460 kW at 8.51 GHz; the signals received by the VLA multi-dish array detected points of radar reflectivity (radar luminosity) with depolarized waves from Mercury's north pole.

Radar maps of the surface of the planet were made using the Arecibo radiotelescope. The survey was conducted with 420 kW UHF band (2.4 GHz) radio waves which allowed for a 15 km resolution. This study not only confirmed the existence of the zones of high reflectivity and depolarization, but also found a number of new areas (bringing the total to 20) and was even able to survey the poles. It has been postulated that surface ice may be responsible for these high luminosity levels, as the silicate rocks that compose most of the surface of Mercury have exactly the opposite effect on luminosity.

In spite of its proximity to the Sun, Mercury may have surface ice, since temperatures near the poles are constantly below freezing point: On the polar plains, the temperature does not rise above −106 °C. And craters at Mercury's higher latitudes (discovered by radar surveys from Earth as well) may be deep enough to shield the ice from direct sunlight. Inside the craters, where there is no solar light, temperatures fall to −171 °C.

Despite sublimation into the vacuum of space, the temperature in the permanently shadowed region is so low that this sublimation is slow enough to potentially preserve deposited ice for billions of years.

At the South Pole, the location of a large zone of high reflectivity coincides with the location of the Chao Meng-Fu crater, and other small craters containing reflective areas have also been identified. At the North Pole, a number of craters smaller than Chao-Meng Fu have these reflective properties.

The strength of the radar reflections seen on Mercury are small compared to that which would occur with pure ice. This may be due to powder deposition that does not cover the surface of the crater completely or other causes, e.g. a thin overlying surface layer. However, the evidence for ice on Mercury is not definitive. The anomalous reflective properties could also be due to the existence of deposits of metallic sulfates or other materials with high reflectance.

Possible origin of ice

Mercury is not unique in having craters that stand in permanent shadow; at the south pole of Earth's Moon there is a large crater (Aitken) where some possible signs of the presence of ice have been seen (although their interpretation is disputed). It is thought by astronomers that ice on both Mercury and the Moon must have originated from external sources, mostly impacting comets. These are known to contain large amounts, or a majority, of ice. It is therefore conceivable for meteorite impacts to have deposited water in the permanently shadow craters, where it would remain unwarmed for possibly billions of years due to the lack of an atmosphere to efficiently conduct heat and stable orientation of Mercury's rotation axis.

Mercury

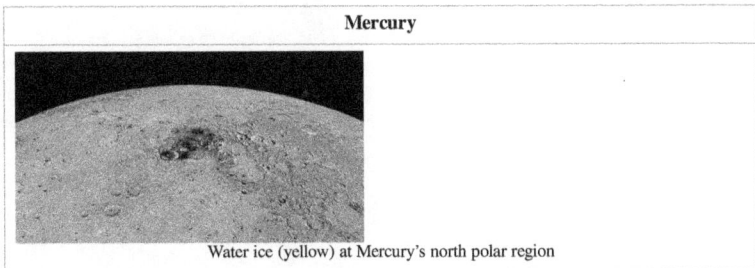

Water ice (yellow) at Mercury's north polar region

References

- *Stardate, Guide to the Solar System*. Publicación de la University of Texas at Austin McDonald Observatory
- *Our Solar System, A Geologic Snapshot*. NASA (NP-157). May 1992.
- Fotografía: *Mercury*. NASA (LG-1997-12478-HQ)
- *This article draws heavily on the corresponding article in the Spanish-language Wikipedia, which was accessed in the version of 26 June 2005.*

Original references for Spanish article

- *Ciencias de la Tierra. Una Introducción a la Geología Física* (*Earth Sciences, an Introduction to Physical Geology*), by Edward J. Tarbuck y Frederick K. Lutgens. Prentice Hall (1999).
- "Hielo en Mercurio" ("Ice on Mercury"). *El Universo, Enciclopedia de la Astronomía y el Espacio* ("The Universe, Encyclopedia of Astronomy and the Space"), Editorial Planeta-De Agostini, p. 141–145. Volume 5. (1997)

External links

- Mariner 10[26]
- MESSENGER probe[27]
- Mercury on Nineplanets.org[28]
- USGS Geology of Mercury[29] Retrieved 5 August 2007

Atmosphere

Atmosphere of Mercury

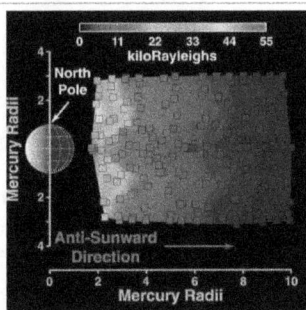

Atmosphere of Mercury[30]

Species	CD,[31] cm^{-2}	SD,[32] cm^{-3}
Hydrogen (H)	$\sim 3 \times 10^9$	~ 250
Molecular hydrogen	$< 3 \times 10^{15}$	$< 1.4 \times 10^7$
Helium	$< 3 \times 10^{11}$	$\sim 6 \times 10^3$
Atomic oxygen	$< 3 \times 10^{11}$	$\sim 4 \times 10^4$
Molecular oxygen	$< 9 \times 10^{14}$	$< 2.5 \times 10^7$
Sodium	$\sim 2 \times 10^{11}$	1.7–3.8 $\times 10^4$
Potassium	$\sim 2 \times 10^9$	~ 4000
Calcium	$\sim 1.1 \times 10^8$	~ 3000
Magnesium	$\sim 4 \times 10^{10}$	$\sim 7.5 \times 10^3$
Argon	$\sim 1.3 \times 10^9$	$< 6.6 \times 10^6$
Water	$< 1 \times 10^{12}$	$< 1.5 \times 10^7$
neon, silicon, sulfur, argon, iron, carbon dioxide, etc.		

Mercury has a very tenuous and highly variable atmosphere (surface-bound exosphere) containing hydrogen, helium, oxygen, sodium, calcium, potassium and water vapor, with a combined pressure level of about 10^{-14} bar (1 nPa). The exospheric species originate either from the Solar wind or from the planetary crust. Solar light pushes the atmospheric gases away from the Sun, creating a comet-like tail behind the planet.

The existence of a Mercurian atmosphere had been contentious before 1974, although by that time a consensus had formed that Mercury, like the Moon, lacked any substantial atmosphere. This conclusion was confirmed in 1974 when the unmanned *Mariner 10* spaceprobe discovered only a tenuous exosphere. Later, in 2008, improved measurements were obtained by the MESSENGER spacecraft, which discovered magnesium in the Mercurian exosphere.

Composition

The Mercurian exosphere consists of a variety of species originating either from the Solar wind or from the planetary crust.[33]   The first constituents discovered were atomic hydrogen (H), helium (He) and atomic oxygen (O), which were observed by the ultraviolet radiation photometer of the *Mariner 10* spaceprobe in 1974. The near-surface concentrations of these elements were estimated to vary from 230 cm^{-3} for hydrogen to 44,000 cm^{-3} for oxygen, with an intermediate concentration of helium. In 2008 the MESSENGER probe confirmed the presence of atomic hydrogen, although its concentration appeared higher than the 1974 estimate.[34] Mercury's exospheric hydrogen and helium are believed to come from the Solar wind, while the oxygen is likely to be of crustal origin.

The fourth species detected in Mercury's exosphere was sodium (Na). It was discovered in 1985 by Drew Potter and Tom Morgan, who observed its Fraunhofer emission lines at 589 and 589.6 nm.[35] The average column density of this element is about 1×10^{11} cm^{-2}. Sodium is observed to concentrate near the poles, forming bright spots.[36] Its abundance is also enhanced near the dawn terminator as compared to the dusk terminator.[37] Some research has claimed a correlation of the sodium abundance with certain surface features such as Caloris or radio bright spots; however these results remain controversial. A year after the sodium discovery, Potter and Morgan reported that potassium (K) is also present in the exosphere of Mercury, though with a column density two orders of magnitude lower than that of sodium. The properties and spatial distribution of these two elements are otherwise very similar.[38] In 1998 another element, calcium (Ca), was detected with column density three orders of

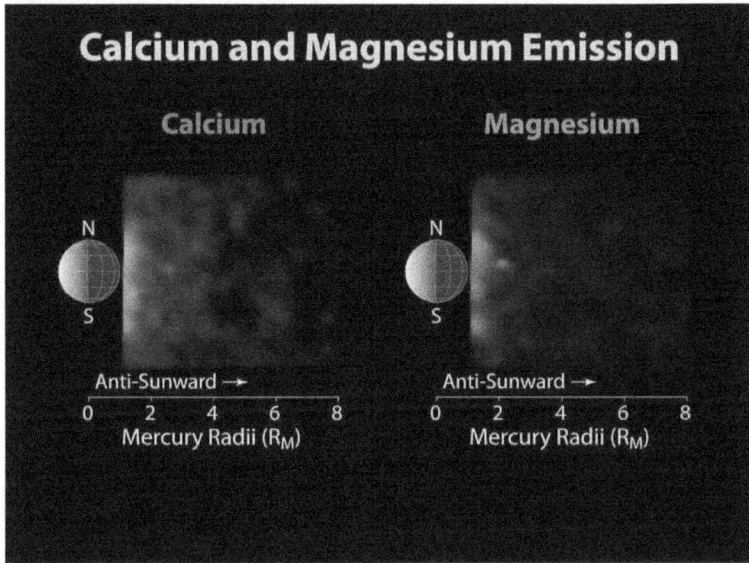

Figure 39: *Ca and Mg in the tail*

magnitude below that of sodium.[39] Observations by the MESSENGER probe in 2009 showed that calcium is concentrated mainly near the equator—opposite to what is observed for sodium and potassium.[40] Further observations by Messenger reported in 2014 note the atmosphere is supplemented by materials vaporized off the surface by meteors both sporadic and in a meteor shower associated with Comet Encke.

In 2008 the MESSENGER probe's Fast Imaging Plasma Spectrometer (FIPS) discovered several molecular and different ions in the vicinity of Mercury, including H_2O^+ (ionized water vapor) and H_2S^+ (ionized hydrogen sulfide). Their abundances relative to sodium are about 0.2 and 0.7, respectively. Other ions such as H_3O^+ (hydroxonium), OH (hydroxyl), O_2^+ and Si^+ are present as well.[41] During its 2009 flyby, the Ultraviolet and Visible Spectrometer (UVVS) channel of the Mercury Atmospheric and Surface Composition Spectrometer (MASCS) on board the MESSENGER spacecraft first revealed the presence of magnesium in the Mercurian exosphere. The near-surface abundance of this newly detected constituent is roughly comparable to that of sodium.

Properties

Mariner 10's ultraviolet observations have established an upper bound on the exospheric surface density at about 10^5 particles per cubic centimeter. This corresponds to a surface pressure of less than 10^{-14} bar (1 nPa).[42]

The temperature of the Mercury's exosphere depends on species as well as geographical location. For exospheric atomic hydrogen, the temperature appears to be about 420 K, a value obtained by both *Mariner 10* and *MESSENGER*. The temperature for sodium is much higher, reaching 750–1500 K on the equator and 1,500–3,500 K at the poles.[43] Some observations show that Mercury is surrounded by a hot corona of calcium atoms with temperature between 12,000 and 20,000 K.

Tails

Because of Mercury's proximity to the Sun, the pressure of solar light is much stronger than near Earth. Solar radiation pushes neutral atoms away from Mercury, creating a comet-like tail behind it.[40] The main component in the tail is sodium, which has been detected beyond 24 million km (1000 R_M) from the planet.[44] This sodium tail expands rapidly to a diameter of about 20,000 km at a distance of 17,500 km.[45] In 2009, *MESSENGER* also detected calcium and magnesium in the tail, although these elements were only observed at distances less than 8 R_M.

References

Bibliography

<templatestyles src="Template:Refbegin/styles.css" />

- Domingue, Deborah L.; Koehn, Patrick L.; Killen, Rosemary M.; et al. (2007). "Mercury's Atmosphere: A Surface-Bounded Exosphere". *Space Science Reviews*. **131** (1-4): 161–186. Bibcode: 2007SSRv..131..161D[46]. doi: 10.1007/s11214-007-9260-9[47].
- Fink, Uwe; Larson, Harold P.; Poppen, Richard F. (1974). "A new upper limit for an atmosphere of CO2, CO on Mercury". *The Astrophysical Journal*. **187**: 407–415. Bibcode: 1967ApJ...149L.137B[48]. doi: 10.1086/180075[49].
- Killen, Rosemary; Cremonese, Gabrielle; Lammer, Helmut; et al. (2007). "Processes that Promote and Deplete the Exosphere of Mercury". *Space Science Reviews*. **132** (2-4): 433–509. Bibcode: 2007SSRv..132..433K[50]. doi: 10.1007/s11214-007-9232-0[51].

- McClintock, William E.; Bradley, E. Todd; Vervack Jr, Ronald J.; et al. (2008). "Mercury's Exosphere: Observations During MESSENGER's First Mercury Flyby". *Science*. **321** (5885): 92–94. Bibcode: 2008Sci... 321...62M[52]. doi: 10.1126/science.1159467[53]. PMID 18599778[54].
- Schmidt, Carl A.; Wilson, Jody K.; Baumgardner, Jeff; Mendillo, Michael (2010). "Orbital effects on Mercury's escaping sodium exosphere". *Icarus*. **207** (1): 9–16. Bibcode: 2010Icar..207....9S[55]. doi: 10.1016/j.icarus.2009.10.017[56].
- McClintock, William E.; Vervack Jr, Ronald J.; Bradley, E. Todd; et al. (2009). "MESSENGER Observations of Mercury's Exosphere: Detection of Magnesium and Distribution of Constituents". *Science*. **324** (5927): 610–613. Bibcode: 2009Sci...324..610M[57]. doi: 10.1126/science.1172525[58]. PMID 19407195[59].
- Rasool, S.I.; Gross, S.H.; McGovern, W.E. (1966). "The atmosphere of Mercury". *Space Science Reviews*. **5** (5): 565–584. Bibcode: 1966SSRv....5..565R[60]. doi: 10.1007/BF00167326[61].
- Williams, I.P. (1974). "Atmosphere of Mercury". *Nature*. **249** (5454): 234. Bibcode: 1974Natur.249..234W[62]. doi: 10.1038/249234a0[63].
- Zurbuchen, Thomas H.; Raines, Jim M.; Gloeckler, George; et al. (2008). "MESSENGER Observations of the Composition of Mercury's Ionized Exosphere and Plasma Environment". *Science*. **321** (5885): 90–92. Bibcode: 2008Sci...321...90Z[64]. doi: 10.1126/science.1159314[65]. PMID 18599777[66].

Magnetic field

Mercury's magnetic field

Magnetosphere of Mercury

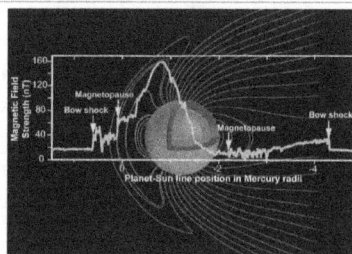

Graph showing relative strength of Mercury's magnetic field.

Discovery	
Discovered by	*Mariner 10*
Discovery date	April 1974
Internal field	
Radius of Mercury	2,439.7 ± 1.0 km
Magnetic moment	2 to 6 × 10^{12} T·m^3
Equatorial field strength	300 nT
Dipole tilt	0.0°
Solar wind parameters	
Speed	400 km/s
Magnetospheric parameters	
Type	Intrinsic
Magnetopause distance	1.4 R_M
Magnetotail length	10–100 R_M

Main ions	Na^+, O^+, K^+, Mg^+, Ca^+, S^+, H_2S^+
Plasma sources	Solar wind
Maximum particle energy	up to 50 keV
Aurora	

Mercury's magnetic field is approximately a magnetic dipole (meaning the field has only two magnetic poles) apparently global, on planet Mercury. Data from *Mariner 10* led to its discovery in 1974; the spacecraft measured the field's strength as 1.1% that of Earth's magnetic field. The origin of the magnetic field can be explained by dynamo theory. The magnetic field is strong enough near the bow shock to slow the solar wind, which induces a magnetosphere.

Strength

The magnetic field is about 1.1% as strong as Earth's. At the Hermean equator, the relative strength of the magnetic field is around 300 nT, which is weaker than that of Jupiter's moon Ganymede. Mercury's magnetic field is weaker than Earth's because its core had cooled and solidified more quickly than Earth's. Although Mercury's magnetic field is much weaker than Earth's magnetic field, it is still strong enough to deflect the solar wind, inducing a magnetosphere. Because Mercury's magnetic field is weak while the interplanetary magnetic field it interacts with in its orbit is relatively strong, the solar wind dynamic pressure at Mercury's orbit is also three times larger than at Earth.

Whether the magnetic field changed to any significant degree between the *Mariner 10* mission and the MESSENGER mission remains an open question. A 1988 J.E.P. Connerney and N.F. Ness review of the Mariner magnetic data noted eight different papers in which were offered no less than fifteen different mathematical models of the magnetic field derived from spherical harmonic analysis of the two close *Mariner 10* flybys, with reported centered magnetic dipole moments ranging from 136 to 350 $nT\text{-}R_M^3$ (R_M is a Mercury radius of 2436 km). In addition they pointed out that "estimates of the dipole obtained from bow shock and/or magnetopause positions (only) range from approximately 200 $nT\text{-}R_M^3$ (Russell 1977) to approximately 400 $nT\text{-}R_M^3$ (Slavin and Holzer 1979b)." They concluded that "the lack of agreement among models is due to fundamental limitations imposed by the spatial distribution of available observations." Anderson *et al.* 2011, using high-quality *MESSENGER* data from many orbits around Mercury – as opposed to just a few high-speed flybys – found that the dipole moment is 195 ± 10 $nT\text{-}R_M^3$.

Figure 40: *Data from Mariner 10 led to the discovery of Mercury's magnetic field.*

Discovery

Before 1974, it was thought that Mercury could not generate a magnetic field because of its relatively small diameter and lack of an atmosphere. However, when *Mariner 10* made a fly-by of Mercury (somewhere around April 1974), it detected a magnetic field that was about 1/100th the total magnitude of Earth's magnetic field. But these passes provided weak constraints on the magnitude of the intrinsic magnetic field, its orientation and its harmonic structure, in part because the coverage of the planetary field was poor and because of the lack of concurrent observations of the solar wind number density and velocity. Since the discovery, Mercury's magnetic field has received a great deal of attention, primarily because of Mercury's small size and slow 59-day-long rotation.

The magnetic field itself is thought to originate from the dynamo mechanism, although this is uncertain as yet.

Origins

The origins of the magnetic field can be explained by the dynamo theory; i.e., by the convection of electrically conductive molten iron in the planet's outer core. A dynamo is generated by a large iron core that has sunk to a planet's center of mass, has not cooled over the years, an outer core that has not been

Figure 41: *Mercury's magnetic field tends to be*
stronger at the equator than at other areas of Mercury.

completely solidified, and circulates around the interior. Before the discovery
of its magnetic field in 1974, it was thought that because of Mercury's small
size, its core had cooled over the years. There are still difficulties with this dynamo theory, including the fact that Mercury has a slow, 59-day-long rotation
that could not have made it possible to generate a magnetic field.

This dynamo is probably weaker than Earth's because it is driven by thermo-
compositional convection associated with inner core solidification. The ther-
mal gradient at the core–mantle boundary is subadiabatic, and hence the outer
region of the liquid core is stably stratified with the dynamo operating only at
depth, where a strong field is generated. Because of the planet's slow rotation,
the resulting magnetic field is dominated by small-scale components that fluc-
tuate quickly with time. Due to the weak internally generated magnetic field
it is also possible that the magnetic field generated by the magnetopause cur-
rents exhibits a negative feedback on the dynamo processes, thereby causing
the total field to weaken.

Magnetic poles and magnetic measurement

Like Earth's, Mercury's magnetic field is tilted, meaning that the magnetic
poles are not located in the same area as the geographic poles. As a result

of the north-south asymmetry in Mercury's internal magnetic field, the geometry of magnetic field lines is different in Mercury's north and south polar regions. In particular, the magnetic "polar cap" where field lines are open to the interplanetary medium is much larger near the south pole. This geometry implies that the south polar region is much more exposed than in the north to charged particles heated and accelerated by solar wind–magnetosphere interactions. The strength of the quadrupole moment and the tilt of the dipole moment are completely unconstrained.

There have been various ways that Mercury's magnetic field has been measured. In general, the inferred equivalent internal dipole field is smaller when estimated on the basis of magnetospheric size and shape ($\sim$150–200 nT R^3). Recent Earth-based radar measurements of Mercury's rotation revealed a slight rocking motion explaining that Mercury's core is at least partially molten, implying that iron "snow" helps maintain the magnetic field. The *MESSENGER* spacecraft will make more than 500 million measurements of Mercury's magnetic field,Wikipedia:Manual of Style/Dates and numbers#Chronological items using its sensitive magnetometer.

Field characteristics

Scientists noted that Mercury's magnetic field can be extremely "leaky," because *MESSENGER* encountered magnetic "tornadoes" during its second fly-by on October 6, 2008, which could possibly replenish the atmosphere (or "exosphere", as referred to by astronomers). When *Mariner 10* made a fly-by of Mercury back in 1974, its signals measured the bow shock, the entrance and exit from the magnetopause, and that the magnetospheric cavity is $\sim$20 times smaller than Earth's, all of which had presumably decayed during the *MESSENGER* flyby. Even though the field is just over 1% as strong as Earth's, its detection by *Mariner 10* was taken by some scientists as an indication that Mercury's outer core was still liquid, or at least partially liquid with iron and possibly other metals.

BepiColombo mission

BepiColombo is a joint mission of the European Space Agency (ESA) and the Japan Aerospace Exploration Agency (JAXA) to Mercury. It is planned to launch in October 2018. Part of its mission objectives will be to elucidate Mercury's magnetic field.

Figure 42: *The MESSENGER spacraft noted that Mercury's magnetic field is responsible for several magnetic "tornadoes" – twisted bundles of magnetic fields connecting the planetary field to interplanetary space – that are some 800 km wide or a third the total radius of the planet.*

Exploration of Mercury

Exploration of Mercury

<templatestyles src="Multiple_image/styles.css" />

The first probe to visit the innermost planet was Mariner 10.

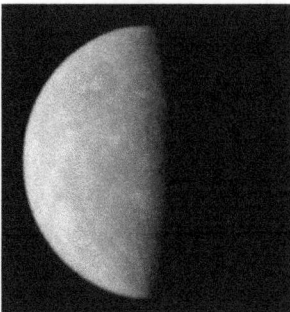

View of Mercury from Mariner 10 in March 1975.

The **exploration of Mercury** has played only a minor role in the space interests of the world. It is the least explored inner planet.[67] As of 2015, the *Mariner 10* and *MESSENGER* missions have been the only missions that have made close observations of Mercury. *MESSENGER* made three flybys before entering orbit around Mercury.[68] A third mission to Mercury, *BepiColombo*, a joint mission between the Japan Aerospace Exploration Agency (JAXA) and the

European Space Agency, is to include two probes. *MESSENGER* and *Bepi-Colombo* are intended to gather complementary data to help scientists understand many of the mysteries discovered by *Mariner 10*'s flybys.

Compared to other planets, Mercury is difficult to explore. The increased speed required to reach it is relatively high, and due to the proximity to the Sun, orbits around it are rather unstable. *MESSENGER* was the first probe to orbit Mercury.

Interest in Mercury

Mercury has not been a primary focus of many space programs. Because the planet is so close to the Sun and spins on its own axis very slowly, its surface temperature varies from 427 °C (801 °F) to –173 °C (–279 °F).[69] The current interest in Mercury is derived from the unexpected observations of *Mariner 10*. Before *Mariner 10*, astronomers thought that the planet simply revolved around the Sun in a highly elliptical orbit. The planet had been observed through ground-based telescopes, and *Mariner 10* provided data that clarified or contradicted many of their inferences.

Another reason why so few missions have targeted Mercury is that it is very difficult to obtain a satellite orbit around the planet on account of its proximity to the Sun, which causes the Sun's gravitational field to pull on any satellite that would be set into Mercury's orbit. The planet's orbit is inclined to the ecliptic by 7°, and its orbital velocity varies from 24.25 miles per second (39.03 km/s) at aphelion to almost 30 miles per second (48 km/s) at perihelion. Spacecraft would be even faster, because they accelerate as they approach the greater gravitational pull of the Sun, but must slow down for orbit insertion, so this entails considerable fuel requirements. This is different with superior planets beyond Earth's orbit where the satellite works against the pull of the Sun. Therefore, reaching an orbit around Mercury requires especially expensive rocketry. Mercury's lack of an atmosphere poses further challenges because it precludes aerobraking. Thus a landing mission would have even more demanding fuel requirements.[70]

Missions

Past missions

Mariner 10

Mariner 10 was a probe whose primary objective was to observe the atmosphere, surface, and physical characteristics of Mercury and Venus. It was a low-cost mission completed for under $98 million.[71] *Mariner 10* was launched at 12:45 am EST on November 3, 1973, from Cape Canaveral.[72] Since Mercury is so close to the Sun it was too difficult to incorporate an orbit around Mercury in the route so *Mariner 10* orbited the Sun. In order to reach its destination, the satellite was accelerated with the gravity field of Venus. It then passed close to Mercury on March 29, 1974, as it flew towards the Sun. This was the first observation made of Mercury at close range. After the encounter *Mariner 10* was in an orbit around the Sun such that for every one of its orbits Mercury made two, and the spacecraft and the planet would be able to meet again. This allowed the probe to pass by Mercury two additional times before completing the mission; these encounters were made on September 21, 1974, and March 16, 1975. However, since the same side of Mercury was illuminated during each of the flybys, at the conclusion of the mission *Mariner 10* had only photographed 45% of its surface. The mission ended when the probe's attitude control gas ran out on March 24, 1975. As the spacecraft was no longer controllable without its nitrogen gas thrusters, a command was sent to the probe to shut down its transmitter.

The close observations collected two important sets of data. The probe detected Mercury's magnetic field, which is very similar to Earth's. This was a surprise to scientists, because Mercury spins so slowly on its axis. Secondly, visual data was provided, which showed the high number of craters on the surface of the planet.[73] The visual data also allowed scientists to determine that Mercury had "not experienced significant crustal modification".[74] This also added to the mystery of the magnetic field, as it was previously believed that the magnetic fields are caused by a molten dynamo effect, but since there was little crustal modification this undermined that idea. The visual data also allowed scientists to investigate the composition and age of the planet.[75]

MESSENGER

MESSENGER was a NASA orbital probe of Mercury. *MESSENGER* stands for "MErcury Surface, Space ENvironment, GEochemistry, and Ranging". It was launched from Cape Canaveral on August 3, 2004, after a one-day delay due to bad weather.[76,77] It took the probe about six and a half years before it entered orbit around Mercury. In order to correct the speed of the satellite it undertook several gravitational slingshot flybys of Earth, Venus and Mercury. It passed

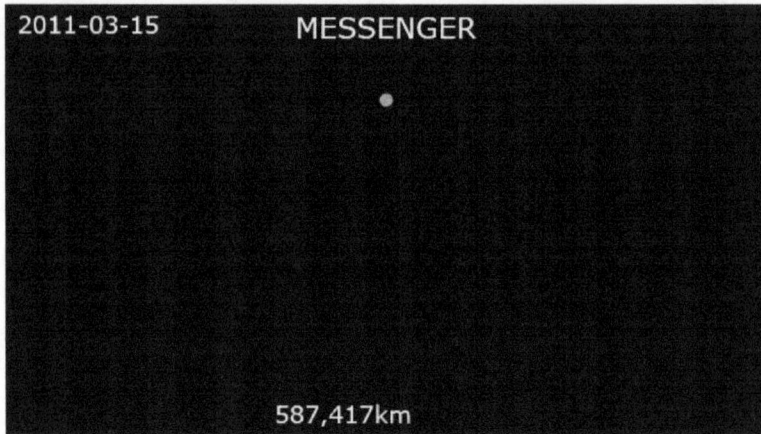

Figure 43: *Animation of MESSENGER's trajectory around*
Mercury from 15 March 2011 to 30 December 2014
MESSENGER · Mercury

by the Earth in February 2005 and then Venus in October 2006 and in October 2007. Furthermore, the probe made three passes of Mercury, one in January 2008, one in October 2008 and one in September 2009, before entering orbit in 2011. During these flybys of Mercury, enough data was collected to produce images of over 95% of its surface.

MESSENGER used a chemical bi-propellant system both to reach Mercury and achieve orbit. *MESSENGER*'s scheduled orbital insertion took place successfully on March 18, 2011. The mission was scheduled to end sometime in 2012, when it was estimated that there would no longer be enough fuel to maintain the probe's orbit.[78] The primary mission was completed on March 17, 2012, having collected close to 100,000 images. *MESSENGER* achieved 100% mapping of Mercury on March 6, 2013, and completed its first year-long extended mission on March 17, 2013. The probe continued collecting scientific data until April 30, 2015, when under a decaying orbit, the probe was allowed to crash onto the surface of Mercury.

The *MESSENGER* mission was designed to study the characteristics and environment of Mercury from orbit. Specifically, the scientific objectives of the mission are:

- characterize the chemical composition of Mercury's surface.
- study the geologic history.
- elucidate the nature of Mercury's magnetic field (magnetosphere).
- determine the size and state of the core.

- determine the volatile inventory at the poles.
- study the nature of Mercury's exosphere.

Future missions

BepiColombo

This mission to Mercury is to include two satellites: the Mercury Planetary Orbiter (MPO) and Mio (Mercury Magnetospheric Orbiter, MMO). Each orbiter has a distinct purpose: the MPO is to acquire images in several wavelengths to map the surface and exosphere composition of Mercury, and Mio's is to study the magnetosphere. The European Space Agency and Japan Aerospace Exploration Agency are working in conjunction on *BepiColombo* and will each provide one of the orbiters. The ESA will provide MPO, while JAXA will provide Mio.[79] The *BepiColombo* mission will attempt to gather enough information to answer these questions:

1. What can we learn from Mercury about the composition of the solar nebula and the formation of the planetary system?
2. Why is Mercury's normalized density markedly higher than that of all other terrestrial planets, as well as the Moon?
3. Is the core of Mercury liquid or solid?
4. Is Mercury tectonically active today?
5. Why does such a small planet possess an intrinsic magnetic field, while Venus, Mars and the Moon do not have any?
6. Why do spectroscopic observations not reveal the presence of any iron, while this element is supposedly the major constituent of Mercury?
7. Do the permanently shadowed craters of the polar regions contain sulfur or water ice?
8. What are the production mechanisms of the exosphere?
9. In the absence of any ionosphere, how does the magnetic field interact with the solar wind?
10. Is Mercury's magnetised environment characterized by features reminiscent of the aurorae, radiation belts and magnetospheric substorms observed on Earth?
11. Since the advance of Mercury's perihelion was explained in terms of space-time curvature, can we take advantage of the proximity of the Sun to test general relativity with improved accuracy?

Like *Mariner 10* and *MESSENGER*, *BepiColombo* will use gravity slingshots from Venus and Earth. *BepiColombo* will use solar electric propulsion (ion engines) and then also use similar manoeuvres at the Moon, Venus, and Mercury. These techniques will slow the orbiters as they approach Mercury. It is essential to avoid using fuel to slow the orbiters as they get closer to the Sun to minimize the gravitational influence of the Sun.

The *BepiColombo* mission was approved in November 2009, with launch date planned for October 2018. It is scheduled to enter orbit around Mercury in February 2024. It will then gather data for one, or possibly two years.

Proposed missions

Mercury-P

Mercury-P (Меркурий-П) is a proposed mission to Mercury by the Russian Space Agency. The currently anticipated launch date is 2031. It is planned to be a lander.

Canceled missions

Mercury Observer

Mercury Observer was a cancelled proposal in the Planetary Observer program.

Comparison of *MESSENGER* and *BepiColombo*

BepiColombo was designed to complement the findings of *MESSENGER* and is equipped with far more measuring equipment than *MESSENGER* to obtain a larger range of data. The orbit patterns of BepiColombo and *MESSENGER* are significantly different.[80]

The MPO will have a circular orbit much closer to Mercury. The reason for this orbit is that the MPO will be measuring the composition of the surface and exosphere, and the close orbit will aid on data quality. On the other hand, the MMO and *MESSENGER* took largely elliptical orbits. This is because of the stability of the orbit and the lower amount of fuel required to obtain and maintain the orbit.[81] Another reason for the different orbits of MMO and *MESSENGER* was to provide complementary data. The data of the two combined satellites will provide more accurate measurements.

External links

- Mariner 10[82]
- *MESSENGER* probe[83]
- Shirley, Donna L. (August 2003). The Mariner 10 mission to Venus and Mercury. *Acta Astronautica*, Aug 2003, Vol. 53, Issue 4-10, p375, 11p; (AN 11471527).

Appendix

References

[1] Extract of page 51 https://books.google.com/books?id=bJoYlBWbCAYC&pg=PA51

[2] Gallant, R. 1986. *The National Geographic Picture Atlas of Our Universe*. National Geographic Society, 2nd edition.

[3] If the area of Washington is about 177 km^2 and 2.5 miles is taken to equal 4 km, Solomon's estimate would equal about 700 cubic kilometres of ice, which would have a mass of about 600 billion tons (6×10^{14} kg).

[4] Mercury Closest Approaches to Earth generated with:
1. Solex 10 http://chemistry.unina.it/~alvitagl/solex/ (Text Output file http://home.surewest.net/kheider/astro/SolexMerc.txt)
2. Gravity Simulator charts http://www.orbitsimulator.com/cgi-bin/yabb/YaBB.pl?num=1235936812
3. JPL Horizons 1950–2200 http://home.surewest.net/kheider/astro/Mercury.txt

[5] U. Le Verrier (1859), (in French), "Lettre de M. Le Verrier à M. Faye sur la théorie de Mercure et sur le mouvement du périhélie de cette planète" https://archive.org/stream/comptesrendusheb49acad#page/378/mode/2up, Comptes rendus hebdomadaires des séances de l'Académie des sciences (Paris), vol. 49 (1859), pp. 379–383. (At p. 383 in the same volume Le Verrier's report is followed by another, from Faye, enthusiastically recommending to astronomers to search for a previously undetected intra-mercurial object.)

[6] (look at 1964 and 2013)

[7] – Numbers generated using the Solar System Dynamics Group, Horizons On-Line Ephemeris System http://ssd.jpl.nasa.gov/horizons.cgi?find_body=1&body_group=mb&sstr=1

[8] See Twilight#Astronomical twilight

[9] , , .

[10] See also the Greek article about the planet.

[11] at pp. 118–122.

[12] Mercury http://scienceworld.wolfram.com/astronomy/Mercury.html at Eric Weisstein's 'World of Astronomy'

[13] Golden, Leslie M., A Microwave Interferometric Study of the Subsurface of the Planet Mercury (1977). PhD Dissertation, University of California, Berkeley

[14] "NASA extends spacecraft's Mercury mission" http://www.upi.com/Top_News/US/2011/11/15/NASA-extends-spacecrafts-Mercury-mission/UPI-55131321408343/. UPI, November 15, 2011. Retrieved November 16, 2011.

[15] https://history.nasa.gov/SP-423/sp423.htm

[16] http://planetarynames.wr.usgs.gov/Page/MERCURY/target

[17] http://planetarynames.wr.usgs.gov/Page/mercuryQuadMap

[18] http://messenger-act.actgate.com/msgr_public_released/react_quickmap.html

[19] https://www.google.com/maps/space/mercury/

[20] http://solarviews.com/eng/mercury.htm

[21] http://www.astronomycast.com/2007/08/episode-49-mercury/

[22] http://messenger.jhuapl.edu/

[23] http://www.esa.int/bepicolombo

[24] Dzurisin D. (1978), *The tectonic and volcanic history of Mercury as inferred from studies of scarps, ridges, troughs, and other lineaments*, Journal of Geophysical Research, v. 83, p. 4883-4906

[25] Van Hoolst, T., Jacobs, C. (2003), *Mercury's tides and interior structure*, Journal of Geophysical Research, v. 108, p. 7.

[26] https://web.archive.org/web/20050905000210/http://www.jpl.nasa.gov/missions/past/mariner10.html

[27] http://messenger.jhuapl.edu/

[28] http://www.nineplanets.org/mercury.html

[29] https://astrogeology.usgs.gov/Projects/BrowseTheGeologicSolarSystem/MercuryGeo.html
[30] Killen 2007, p. 456, Table 5
[31] Column density
[32] Surface density
[33] Killen, 2007, pp. 433–434
[34] McClintock 2008, p. 93
[35] Killen, 2007, pp. 434–436
[36] Killen, 2007, pp. 438–442
[37] Killen, 2007, pp. 442–444
[38] Killen, 2007, pp. 449–452
[39] Killen, 2007, pp. 452–453
[40] McClintock 2009, p. 612–613
[41] Zurbuchen 2008, p. 91, Table 1
[42] Domingue, 2007, pp. 162–163
[43] Killen, 2007, pp. 436–438
[44] Schmidt 2010, p. 9–16
[45] Killen, 2007, p. 448
[46] http://adsabs.harvard.edu/abs/2007SSRv..131..161D
[47] //doi.org/10.1007/s11214-007-9260-9
[48] http://adsabs.harvard.edu/abs/1967ApJ...149L.137B
[49] //doi.org/10.1086/180075
[50] http://adsabs.harvard.edu/abs/2007SSRv..132..433K
[51] //doi.org/10.1007/s11214-007-9232-0
[52] http://adsabs.harvard.edu/abs/2008Sci...321...62M
[53] //doi.org/10.1126/science.1159467
[54] //www.ncbi.nlm.nih.gov/pubmed/18599778
[55] http://adsabs.harvard.edu/abs/2010Icar..207....9S
[56] //doi.org/10.1016/j.icarus.2009.10.017
[57] http://adsabs.harvard.edu/abs/2009Sci...324..610M
[58] //doi.org/10.1126/science.1172525
[59] //www.ncbi.nlm.nih.gov/pubmed/19407195
[60] http://adsabs.harvard.edu/abs/1966SSRv....5..565R
[61] //doi.org/10.1007/BF00167326
[62] http://adsabs.harvard.edu/abs/1974Natur.249..234W
[63] //doi.org/10.1038/249234a0
[64] http://adsabs.harvard.edu/abs/2008Sci...321...90Z
[65] //doi.org/10.1126/science.1159314
[66] //www.ncbi.nlm.nih.gov/pubmed/18599777
[67] JHU/APL (2006). MESSENGER: MErcury Surface, Space ENvironment, GEochemistry, and Ranging http://messenger.jhuapl.edu/ Retrieved on 2007-01-27
[68] Munsell Kirk-editor (November 06, 2006). NASA: Solar System Exploration: Missions to Mercury http://solarsystem.jpl.nasa.gov/missions/profile.cfm?Sort=Target&Target=Mercury&MCode=MESSENGER. Retrieved on 2007-01-27.
[69] Munsell Kirk-editor (November 06, 2006). NASA: Solar System Exploration: Planet Mercury http://solarsystem.jpl.nasa.gov/planets/profile.cfm?Object=Mercury&Display=Overview. Retrieved on 2007-01-27.
[70] http://www.esa.int/For_Media/Press_Releases/Critical_decisions_on_Cosmic_Vision
[71] Shirley, 2003
[72] Dunne, James A. (1978). The Voyage of Mariner 10: Mission to Venus and Mercury (NASA SP-424). U.S. Government Printing Office. p. 45. ASIN B000C19QHA.
[73] Dunne, 1978, p. 74
[74] Dunne, 1978, p. 101
[75] Dunne, 1978, p. 103
[76] Malik, T. (2004). Mercury MESSENGER Launch Postponed http://www.space.com/206-mercury-messenger-launch-postponed-tuesday.html. Retrieved 2015-07-18.

[77] NBC News (2004). NASA launches spacecraft to Mercury http://www.nbcnews.com/id/5577224/ns/technology_and_science-space/t/nasa-launches-spacecraft-mercury/#.VaqRNvnvngA. Retrieved 2015-07-18.

[78] Planetary Society(2007) Space Topics: *MESSENGER* http://planetary.org/explore/topics/messenger/. Retrieved 11/9/2010

[79] ESA (2007). BepiColombo http://sci.esa.int/science-e/www/area/index.cfm?fareaid=30. Retrieved 2007-02-01.

[80]

[81] Mukai, T.; Yamakawa, H.; Hayakawa, H.; Kasaba, Y.; and Ogawa, H (2006). Present status of the BepiColombo/Mercury magnetospheric orbiter http://www.sciencedirect.com/science/article/pii/S0273117705011579. *Advances in Space Research*, Volume 38, Issue 4, Mercury, Mars and Saturn, 2006, Pages 578-582.

[82] https://web.archive.org/web/20050905000210/http://www.jpl.nasa.gov/missions/past/mariner10.html

[83] http://messenger.jhuapl.edu/

Article Sources and Contributors

The sources listed for each article provide more detailed licensing information including the copyright status, the copyright owner, and the license conditions.

Mercury (planet) *Source:* https://en.wikipedia.org/w/index.php?oldid=855734850 *License:* Creative Commons Attribution-Share Alike 3.0 *Contributors:* 331dot, A2soup, Abatmose, Alaney2k, Alexander Davronov, Alexandritechrysoberyl, Andyjsmith, Arado, Archon 2488, Arianewiki1, Aveh8, BD2412, Bagunceiro, BatteryIncluded, Becky Sayles, Begoon, Beland, Bencherlite, Bender235, Bomb319, Bumm13, Cablehorn, CarolynMW, ComicsAre-JustAllRight, Crisco 1492, CuriousEric, DN-boards1, DOwenWilliams, DVdm, Dan Gluck, Dawnseeker2000, DePiep, Deeday-UK, DelftUser, Dhtwiki, Dlthewave, DocWatson42, Double sharp, DrKay, Drbogdan, Dutchy45, Earthandmoon, Edulovers, Efroimsk, Ehrenkater, FlightTime, GB fan, GeoffreyT2000, Geographyinitiative, Gnana Sreekar, Hedwig in Washington, Hellbus, Hesperian, Hillbillyholiday, Hms1103, Hoho24, Huntster, Huritisho, Iggy the Swan, IggyKoopa408, Jcpag2012, JeanLucMargot, Jim.henderson, John, John85, Jon Kolbert, JorisvS, Jpgordon, Kharkiv07, Kheider, Koavf, Kohran, Kudzu1, Kwamikagami, L3erdnik, Lollipop, Loraof, LouScheffer, Maczkopeti, MaelstromOfSilence, Manubhatt3, MartinZ, Materialscientist, Mc-donalds, Mikhail Ryazanov, Mitchumch, Modest Genius, Mousenight, Mr Stephen, Mx. Granger, NSH002, Nedim Ardoğa, Nihiltres, Nijoakim, Nvvchar, Nwbeeson, Oldag07, Paleontologist99, Pawyilee, Pdebee, PhilipTerryGraham, Piledhigheranddeeper, PlanetUser, Plastikspork, PlayStation 14, Primefac, Purgy Purgatorio, QuartierLatin1968, Rfassbind, Rjwilmsi, Rodrigolopes, Rothorpe, Rpot2, Rusl, Sailsbystars, Saros136, Sbmeirow, Schuy B., Signedzzz, SkoreKeep, Slightsmile, Smk65536, Smkolins, Spacemarine10, Steve03Mills, SucreRouge, Sumanth699, TAnthony, TJRC, Tetra quark, Thanatos666, The Transhumanist, TheGlatiator, TheWhistleGag, Thecodingproject, Thor Dockweiler, Thryduulf, Tom.Reding, Trappist the monk, Twinsday, Tóraf, Ugly Ketchup, Urhixidur, Velho, VirtualDave, Voello, W like wiki, Waldir, White whirlwind, Wieralee, Wikifan2744, WolfmanSF, WurmWoode, YBG, Zundark 1

Geology of Mercury *Source:* https://en.wikipedia.org/w/index.php?oldid=846164259 *License:* Creative Commons Attribution-Share Alike 3.0 *Contributors:* A Train, Agastya, Alias, Amayarora, Andy120290, AshLin, BD2412, Backpackadam, Benbest, Bobblewik, Bryan Derksen, Chmee2, Chris the speller, ChrisGualtieri, ClueBot NG, Cmapm, Craxyxarc, Cyclopia, DMacks, DOSGuy, DaveGorman, Dawnseeker2000, Deuar, Doovinator, Draeco, Drbogdan, Dual Freq, Etacar11, Fabio Bettani, Fenice, FoCuSandLeArN, Gamemasterv, GeoWriter, Giannioneill, Hajor, Halidecyphon, Hedwig in Washington, Hesperian, I dream of horses, Icairns, J3ff, Jcpag2012, Jmabel, Joelholdsworth, Joldy, JorisvS, K Lepo, Kaldari, Kentholke, KnightRider~enwiki, Knowledge Seeker, Kwamikagami, LAAFan, Lightmouse, MTSbot~enwiki, Matsmbal, Matthewprc, Maxim, Megalodon99, Mikenorton, Muhandes, Nickptar, Originalwana, Panellet, Pilaf~enwiki, Ptbotgourou, Qwarc, R'n'B, RandyUang, Remember, Reyk, Rich Farmbrough, Ricnun, Rjwilmsi, Roentgenium111, Romanm, Roy da Vinci, Sardanaphalus, SchreiberBike, Sietse Snel, SkyWalker, Solar-Wind, Sorsanmetsastaja, Stan2525, Tom.Reding, Tycho, Ultraexactzz, Unused007, Vieque, Wdfarmer, WolfmanSF, Zxcvbnm, Zykure, さえぼー, 57 anonymous edits 37

Atmosphere of Mercury *Source:* https://en.wikipedia.org/w/index.php?oldid=829130537 *License:* Creative Commons Attribution-Share Alike 3.0 *Contributors:* $arahxyz, 84user, APT, Alansohn, Andonic, Andy120290, Anonymous Dissident, Auric, Bender235, Beyond My Ken, Calmer Waters, CambridgeBayWeather, Cbrick77, Cgingold, Chmee2, ClueBot NG, Colchicum, Courcelles, DMacks, Diannaa, DisillusionedBitterAndKnackered, Dthomsen8, Edulovers, Egmontaz, Entropy, EoGuy, Excirial, Gene Nygaard, Giftlite, Headbomb, Icairns, Idempotent, Intelligentsium, J 1982, JamesBWatson, JorisvS, Josve05a, Karthikeyan K200878, Khanadarhodes, Kitch, KuduIO, Lanthanum-138, Liam McM, Lil'icarus, Mild Bill Hiccup, MusikAnimal, NewEnglandYankee, ObsidianComet, Owllord97, PST, Pamdhiga, Philip Trueman, Piano non troppo, RJHall, Racerx11, Radon210, Rbethune, Rds719, Reatlas, Remember, Rich Farmbrough, Ricnun, Rursus, Ruslik0, Sardanaphalus, SchuminWeb, ScottSteiner, Scutigera, Seaphoto, Slon02, Smalljim, Smithbrenon, Smkolins, Soojus, StealthCopyEditor, Steel1943, Stephenb, Stickee, The Obento Musubi, Thirdright, Tide rolls, ToBeFree, Tom.Reding, Trappist the monk, Trifoliate, Trollpediahehe, VoABot II, 120 anonymous edits 51

Mercury's magnetic field *Source:* https://en.wikipedia.org/w/index.php?oldid=844849369 *License:* Creative Commons Attribution-Share Alike 3.0 *Contributors:* Anthony Appleyard, Anypodetos, Barak Thunder, Bruce1ee, Carbon6, ClueBot NG, Double sharp, Eric-Wester, Gap9551, GoingBatty, Headbomb, Improbable llama, Isambard Kingdom, Jim.henderson, JorisvS, Lchiarav, Lesabendio2, Materialscientist, Reatlas, Rjwilmsi, RockMagnetist, Shaun, Sietse Snel, SpaceChimp1992, Steel1943, ThatRusskiiGuy, Tom.Reding, Vultur~enwiki, Whoop whoop pull up, 14 anonymous edits 57

Exploration of Mercury *Source:* https://en.wikipedia.org/w/index.php?oldid=852917438 *License:* Creative Commons Attribution-Share Alike 3.0 *Contributors:* A5b, ACSE, Altamel, Andy120290, AnthonyHolmes, Arado, AssegaiAli, BatteryIncluded, Bility, Blathnaid, Brian A Schmidt, Brian the Editor, Bryan Derksen, Brynnium, Canopus27, ChiZeroOne, ClueBot NG, Cnwilliams, Colonel Wilhelm Klink, Czolgolz, DMacks, DeDutch72, De-Noel, DiscordantNote, Donner60, Elee, Element16, EvergreenFir, Fanyavizuri, Funandtrvl, Ginsuloft, Graywords, HMSLavender, Haruyasha, Hazard-SJ, Headbomb, Hillbillyholiday, Hms1103, JFG, JJMC89, JSheff, Jim.henderson, Jim1138, JorisvS, Josve05a, Kaystay, Keith D, Kitty1976, Kmg90, Leor klier, Lightmouse, Lordzinga, Lugia2453, LynxTufts, Materialscientist, Meters, Mevagiss, Mlm42, Naraht, Narky Blert, Ninney, Pascal.Tesson, Paul Pot, Phoenix7777, Poolio, PresidentCooper, Randy Kryn, Reatlas, Recognizance, Remember, Rothorpe, Rreagan007, Ruslik0, Sardanaphalus, Scarian, SchuminWeb, Sciurinæ, Sesu Prime, Sizeofint, Smart30, SoxBot III, StevenBjerke, Stone, Suliivja, Tetra quark, The Quixotic Potato, TheCatalyst31, TheCodeman4, Thingg, Tom.Reding, Tony Mach, Unbuttered Parsnip, Vsmith, Vuerqex, Wbm1058, Williamhhortner, Wolfkeeper, Yill577, Ylee, 113 anonymous edits 63

Image Sources, Licenses and Contributors

The sources listed for each image provide more detailed licensing information including the copyright status, the copyright owner, and the license conditions.

Image *Source:* https://en.wikipedia.org/w/index.php?title=File:Padlock-silver.svg *Contributors:* AzaToth, BotMultichill, BotMultichillT, Gurch, Jarekt, Kallerna, Multichill, Perhelion, Rd232, Riana, Sarang, Siebrand, Steinsplitter, 4 anonymous edits 1

Image *Source:* https://en.wikipedia.org/w/index.php?title=File:Cscr-featured.svg *License:* GNU Lesser General Public License *Contributors:* Anomie 1

Image *Source:* https://en.wikipedia.org/w/index.php?title=File:Mercury_in_color_-_Prockter07-edit1.jpg *License:* Public Domain *Contributors:* NASA/Johns Hopkins University Applied Physics Laboratory/Carnegie Institution of Washington 1

Image *Source:* https://en.wikipedia.org/w/index.php?title=File:Loudspeaker.svg *License:* Public Domain *Contributors:* User:Dbenbenn, User:Optimager, User:Tsca, User:Dbenbenn, User:Optimager, User:Tsca, User:Dbenbenn, User:Optimager, User:Tsca 1

Figure 1 *Source:* https://en.wikipedia.org/w/index.php?title=File:Internal_Structure_of_Mercury.jpg *License:* Public Domain *Contributors:* User:Jcpag2012 4

Figure 2 *Source:* https://en.wikipedia.org/w/index.php?title=File:Gravity_Anomalies_on_Mercury.jpg *License:* Public Domain *Contributors:* NASA/Goddard Space Flight Center Science Visualization Studio/Johns Hopkins University Applied Physics Laboratory/Carneg 5

Image *Source:* https://en.wikipedia.org/w/index.php?title=File:PIA19420-Mercury-NorthHem-Topography-MLA-Messenger-20150416.jpg *License:* Public Domain *Contributors:* Drbogdan, Illustr, Lotse 7

Figure 3 *Source:* https://en.wikipedia.org/w/index.php?title=File:EW1027346412Gnomap.png *License:* Public Domain *Contributors:* NASA/Johns Hopkins University Applied Physics Laboratory/Carnegie Institution of Washington 8

Figure 4 *Source:* https://en.wikipedia.org/w/index.php?title=File:Unmasking_the_Secrets_of_Mercury.jpg *License:* Public Domain *Contributors:* PlanetUser, Rhadamante 8

Figure 5 *Source:* https://en.wikipedia.org/w/index.php?title=File:PIA19450-PlanetMercury-CalorisBasin-20150501.jpg *License:* Public Domain *Contributors:* Drbogdan 9

Figure 6 *Source:* https://en.wikipedia.org/w/index.php?title=File:PIA19421-Mercury-Craters-MunchSanderPoe-20150416.jpg *License:* Public Domain *Contributors:* Drbogdan, Lotse 10

Figure 7 *Source:* https://en.wikipedia.org/w/index.php?title=File:PIA19423-Mercury-AbedinCrater-20150416.jpg *License:* Public Domain *Contributors:* Drbogdan, Lotse 11

Figure 8 *Source:* https://en.wikipedia.org/w/index.php?title=File:1_Denevi_5_Degas_crater_MESSENGER_soacecraft.jpg *License:* Public Domain *Contributors:* Pline, Terriffic Dunker Guy 11

Image *Source:* https://en.wikipedia.org/w/index.php?title=File:Mercury_weird_terrain.jpg *License:* Public Domain *Contributors:* NASA/JPL .. 11

Figure 9 *Source:* https://en.wikipedia.org/w/index.php?title=File:Picasso_crater.png *License:* Public Domain *Contributors:* NASA/Johns Hopkins University Applied Physics Laboratory/Carnegie Institution of Washington 13

Figure 10 *Source:* https://en.wikipedia.org/w/index.php?title=File:Mercury_Globe-MESSENGER_mosaic_centered_at_0degN-0degE.jpg *License:* Public Domain *Contributors:* Illustr, Jcpag2012, Jianhui67, Tdadamemd, Trijnstel, 5 anonymous edits 14

Figure 11 *Source:* https://en.wikipedia.org/w/index.php?title=File:Merc_fig2sm.jpg *License:* Public Domain *Contributors:* Quote from : "NASA photo by..." 14

Figure 12 *Source:* https://en.wikipedia.org/w/index.php?title=File:North_pole_of_Mercury_-_-_NASA.jpg *License:* Public Domain *Contributors:* NASA 15

Figure 13 *Source:* https://en.wikipedia.org/w/index.php?title=File:Mercury_Magnetic_Field_NASA.jpg *License:* Public Domain *Contributors:* Joancreus, Lotse, Materialscientist, Ruslik0, Soerfm 16

Image *Source:* https://en.wikipedia.org/w/index.php?title=File:ThePlanets_Orbits_Mercury_PolarView.svg *License:* Creative Commons Attribution-Sharealike 2.5 *Contributors:* User:Eurocommuter 17

Image *Source:* https://en.wikipedia.org/w/index.php?title=File:Mercuryorbitsolarsystem.gif *License:* Creative Commons Attribution-Sharealike 3.0 *Contributors:* User:Lookang 18

Figure 14 *Source:* https://en.wikipedia.org/w/index.php?title=File:Mercury's_orbital_resonance.svg *License:* GNU Free Documentation License *Contributors:* Tos, converted to SVG from PNG. Original author: Worldtraveller 20

Figure 15 *Source:* https://en.wikipedia.org/w/index.php?title=File:Mercury.jpg *License:* Public Domain *Contributors:* Bricktop, Bryan Derksen, Dbenbenn, PlanetUser, Ruslik0, Sobi3ch~commonswiki, Spundun 22

Figure 16 *Source:* https://en.wikipedia.org/w/index.php?title=File:PIA19247-Mercury-NPolarRegion-Messenger20150316.jpg *License:* Public Domain *Contributors:* Drbogdan, Illustr, JorisvS, Lotse, Terriffic Dunker Guy 23

Figure 17 *Source:* https://en.wikipedia.org/w/index.php?title=File:PIA19422-Mercury-CarnegieRupes-MDIS-MLA-20150416.jpg *License:* Public Domain *Contributors:* Drbogdan, Lotse 24

Figure 18 *Source:* https://en.wikipedia.org/w/index.php?title=File:Mercury-bonatti.png *License:* Public Domain *Contributors:* AtelierMonpli, Lotse, Smerdis of Tlön, Torsch 26

Figure 19 *Source:* https://en.wikipedia.org/w/index.php?title=File:Shatir500.jpg *License:* Public Domain *Contributors:* Azdi80, Konstable~commonswiki, Leinad-Z~commonswiki, Soerfm, 1 anonymous edits 27

Figure 20 *Source:* https://en.wikipedia.org/w/index.php?title=File:Transit_Of_Mercury,_May_9th,_2016.png *Contributors:* User:Elijah.mathews 28

Figure 21 *Source:* https://en.wikipedia.org/w/index.php?title=File:Planet_Elongation.jpg *License:* Public Domain *Contributors:* NASA 28

Figure 22 *Source:* https://en.wikipedia.org/w/index.php?title=File:PIA19411-Mercury-WaterIce-Radar-MDIS-Messenger-20150416.jpg *License:* Public Domain *Contributors:* Drbogdan, Lotse, Terriffic Dunker Guy 30

Figure 23 *Source:* https://en.wikipedia.org/w/index.php?title=File:MESSENGER_Assembly.jpg *License:* Public Domain *Contributors:* NASA 31

Figure 24 *Source:* https://en.wikipedia.org/w/index.php?title=File:PIA18389-MarsCuriosityRover-MercuryTransitsSun-20140603.gif *License:* Public Domain *Contributors:* Drbogdan, Huntster, Lotse, PhilipTerryGraham, Romkur, Wieralee, 23 anonymous edits 31

Figure 25 *Source:* https://en.wikipedia.org/w/index.php?title=File:Mariner_10.jpg *License:* Public Domain *Contributors:* NASA 32

Figure 26 *Source:* https://en.wikipedia.org/w/index.php?title=File:Details_of_MESSENGER's_Impact_Location.jpg *License:* Public Domain *Contributors:* Jcpag2012, Túrelio 33

Figure 27 *Source:* https://en.wikipedia.org/w/index.php?title=File:PIA19449-PlanetMercury-MESSENGER-Images-First-20110329-Last-20150430.jpg *License:* Public Domain *Contributors:* Drbogdan, Lotse, PlanetUser 34

Image *Source:* https://en.wikipedia.org/w/index.php?title=File:Mercury,_Earth_size_comparison.jpg *License:* Public Domain *Contributors:* Jcpag2012, JorisvS, Lotse 35

Image *Source:* https://en.wikipedia.org/w/index.php?title=File:Terrestrial_planet_sizes.jpg *License:* Public Domain *Contributors:* Jcpag2012, Lotse, Wieralee 35

Image *Source:* https://en.wikipedia.org/w/index.php?title=File:1e6m_comparison_Mars_Mercury_Moon_Pluto_Haumea_-_no_transparency.png *Contributors:* Paul Stansifer, 84user, NASA, Celestia, JPL/Caltech 35

Image *Source:* https://en.wikipedia.org/w/index.php?title=File:Sound-icon.svg *License:* GNU Lesser General Public License *Contributors:* Crystal SVG icon set 35

Figure 28 *Source:* https://en.wikipedia.org/w/index.php?title=File:A_Patch_of_Black_On_Mercury.jpg *License:* Public Domain *Contributors:* NASA/Johns Hopkins University Applied Physics Laboratory/Carnegie Institution of Washington 38

Figure 29 *Source:* https://en.wikipedia.org/w/index.php?title=File:Mercury_Double-Ring_Impact_Basin.png *License:* Public Domain *Contributors:* NASA/Johns Hopkins University Applied Physics Laboratory/Carnegie Institution of Washington 38

Figure 30 *Source:* https://en.wikipedia.org/w/index.php?title=File:Mariner_10.jpg *License:* Public Domain *Contributors:* NASA 39

Figure 31 *Source:* https://en.wikipedia.org/w/index.php?title=File:Internal_Structure_of_Mercury.jpg *License:* Public Domain *Contributors:* User:Jcpag2012 40

Figure 32 *Source:* https://en.wikipedia.org/w/index.php?title=File:Gravity_Anomalies_on_Mercury.jpg *License:* Public Domain *Contributors:* NASA/Goddard Space Flight Center Science Visualization Studio/Johns Hopkins University Applied Physics Laboratory/Carneg 41

Image *Source:* https://en.wikipedia.org/w/index.php?title=File:Wikinews-logo.svg *License:* Creative Commons Attribution-Sharealike 3.0 *Contributors:* Vectorized by Simon 01:05, 2 August 2006 (UTC) Updated by Time3000 17 April 2007 to use official Wikinews colours and ap 42

Figure 33 *Source:* https://en.wikipedia.org/w/index.php?title=File:Caloris_basin_labeled.png *License:* Public Domain *Contributors:* NASA .. 44

Figure 34 *Source:* https://en.wikipedia.org/w/index.php?title=File:Unmasking_the_Secrets_of_Mercury.jpg *License:* Public Domain *Contributors:* PlanetUser, Rhadamante 44

License

Index

www.ingramcontent.com/pod-product-compliance
Lightning Source LLC
Chambersburg PA
CBHW022049190326
41520CB00008B/748